Urban Heat Island Mitigation: A Comprehensive Guide to Cool Pavements, Sustainable Cities, and Climate Change Adaptation

Chapter 1: Introduction

Chapter 2: Understanding Urban Heat Islands

Chapter 3: Cool Pavements: An Effective Solution

Chapter 4: Design and Implementation of Cool Pavements

Chapter 5: Integrating Cool Pavements into Urban Planning

Chapter 6: Climate Change Adaptation through Cool Pavements

Chapter 7: Sustainable Cities and Cool Pavements

Chapter 8: Challenges and Barriers to Implementation

Chapter 9: Future Trends and Innovations

Chapter 10: Conclusion

Chapter 1: Introduction

Urban Heat Island (UHI) effect is an increasingly significant environmental challenge faced by cities around the world. As urban areas expand and develop, they experience higher temperatures than their rural surroundings, creating urban heat islands. This phenomenon not only exacerbates the impacts of climate change but also poses serious health, environmental, and economic challenges. Addressing the UHI effect is critical for the sustainability and resilience of urban environments.

This book, " Urban Heat Island Mitigation: A Comprehensive Guide to Cool Pavements, Sustainable Cities, and Climate Change Adaptation," is designed to provide a comprehensive overview of the UHI effect and explore the potential of cool pavements as a key mitigation strategy. Cool pavements, with their ability to reflect more sunlight and absorb less heat than traditional paving materials, offer a practical and effective solution to reduce urban temperatures and enhance the quality of life in cities.

The guide begins with an in-depth examination of the UHI effect, exploring its causes, impacts, and global significance. It then delves into cool pavements, detailing their types, benefits, and the materials and technologies used in their construction. Practical aspects of planning, designing, and implementing cool pavements are thoroughly covered and supported by case studies from various cities.

In addition to the technical aspects, this book highlights the importance of integrating cool pavements into broader urban planning and policy frameworks. It emphasizes the role of cool pavements in climate change adaptation and urban sustainability, showcasing how they can be a part of a holistic approach to creating resilient cities. The guide also addresses the challenges and barriers to implementing cool pavements and explores future trends and innovations in the field.

The primary objective of this book is to equip urban planners, policymakers, engineers, and environmental professionals with the knowledge and tools necessary to effectively mitigate the UHI effect. By fostering a multidisciplinary approach, this guide encourages collaboration across various sectors to implement cool pavement projects and promote sustainable urban development.

Whether you are a seasoned professional or a student of urban planning and environmental science, this book offers valuable insights and practical guidance on addressing one of the most pressing environmental issues of our time. By understanding and applying the concepts and strategies discussed in this guide, you can create cooler, healthier, and more sustainable urban environments.

Welcome to " Urban Heat Island Mitigation: A Comprehensive Guide to Cool Pavements, Sustainable Cities, and Climate Change Adaptation." Let us embark on this journey towards a cooler, more resilient future for our cities.

Urban Heat Island Effect: An Overview

The UHI effect, a significant and growing environmental challenge, causes urban areas to experience considerably higher temperatures than their rural surroundings due to human activities and land use changes.

Definition of UHI

The UHI effect occurs when urban areas experience higher temperatures than their rural surroundings. This temperature difference arises because natural land cover, such as vegetation, is replaced by surfaces like concrete, asphalt, and buildings that absorb and retain heat. During the day, these urban materials store heat from the sun, and at night, they release it slowly, resulting in higher temperatures that persist throughout the evening. This localized warming effect is most pronounced in densely built cities with limited green spaces.

Historical Context and Development

The concept of the UHI effect has been recognized for over two centuries. The first documented observation was by Luke Howard, an English meteorologist, in the early 19th century. In 1818, Howard noted that temperatures in London were consistently higher than in the surrounding countryside. This observation laid the groundwork for the scientific understanding of UHI.

As cities grew and industrialized during the 19th and 20th centuries, the UHI effect became more pronounced. The widespread use of materials like concrete and asphalt, along with the reduction of vegetation, intensified the heat retention in urban areas. The advent of air conditioning in the mid-20th century further contributed to the problem, as the heat expelled from cooling systems added to the urban temperature burden.

With the rise of urbanization, particularly in the latter half of the 20th century, researchers began to study the UHI effect more systematically. They observed that cities with rapid population growth and industrial activity experienced significantly higher ambient temperatures than rural areas. The development of satellite technology in the 1970s allowed for more precise measurement and mapping of UHI effects across various cities globally.

Global Significance of UHI

The UHI effect has far-reaching implications for urban living and the environment. On a global scale, UHI contributes to the planet's overall warming, exacerbating the effects of climate change. Elevated temperatures in cities can amplify heat waves, making urban areas more vulnerable to extreme heat events. This is particularly concerning as the frequency and intensity of heat waves are projected to increase with global climate change.

The UHI effect poses significant risks to public health. Higher temperatures in urban areas can lead to increased incidences of heat-related illnesses and fatalities, especially among vulnerable populations such as the elderly, children, and those with pre-existing health conditions. As heat stress, heat exhaustion, and heatstroke become more prevalent, healthcare systems and public health resources are strained.

The environmental impact of UHI extends beyond temperature increases. It can alter local weather patterns, reduce air quality, and negatively affect urban biodiversity. For example, higher temperatures can increase the formation of ground-level ozone, a harmful air pollutant. Additionally, elevated temperatures can stress urban vegetation and water bodies, disrupting local ecosystems and reducing biodiversity.

Economically, the UHI effect imposes considerable costs on cities. Increased energy consumption for cooling buildings leads to higher utility bills for residents and businesses. This heightened electricity demand often results in greater emissions of greenhouse gases, further contributing to global warming. Moreover, the need for enhanced cooling infrastructure, such as air conditioning systems and heat-resistant materials, represents a significant financial burden for urban areas.

Importance of Mitigating UHI

Mitigating the UHI effect is crucial for safeguarding public health, protecting the environment, and reducing economic costs associated with increased urban temperatures. The UHI effect intensifies the heat burden in cities, leading to a multitude of adverse impacts that necessitate urgent and comprehensive mitigation strategies.

Health Impacts

One of the most pressing reasons to mitigate the UHI effect is its significant impact on public health. Elevated temperatures in urban areas can lead to heat-related illnesses, including heat stress, heat exhaustion, and heatstroke. These conditions are particularly dangerous for vulnerable populations such as the elderly, children, and individuals with pre-existing health conditions. During heatwaves, exacerbating these health issues can result in increased hospital admissions and, in severe cases, fatalities.

Furthermore, higher urban temperatures contribute to the formation of ground-level ozone, a harmful air pollutant. Increased ozone levels can aggravate respiratory conditions like asthma and chronic obstructive pulmonary disease (COPD), further endangering public health. The combined effects of heat and air pollution severely burden healthcare systems, requiring more resources and increasing healthcare costs.

Environmental Impacts

The environmental impacts of the UHI effect extend beyond just elevated temperatures. One of the most significant consequences is the increased demand for energy, particularly for air conditioning. This heightened energy consumption increases greenhouse gas emissions, contributing to global climate change. As cities consume more energy to cool buildings, they inadvertently exacerbate the problem they are trying to mitigate.

In addition to energy consumption, the UHI effect can alter local weather patterns and reduce air quality. Higher temperatures can increase the prevalence of heatwaves, making urban areas more susceptible to extreme weather events. The stress on urban vegetation and water bodies due to elevated temperatures can disrupt local ecosystems, reducing biodiversity and impairing the ecological health of urban environments.

The heat retained by urban surfaces also affects water quality. For example, warmer temperatures can increase the thermal pollution of water bodies, affecting aquatic life and the overall health of urban waterways. This, in turn, can impact the availability and quality of water resources for human consumption and other uses.

Economic Impacts

From an economic perspective, the UHI effect imposes substantial costs on cities and their inhabitants. Increased energy consumption for cooling leads to higher utility bills for residents and businesses. This financial burden is particularly challenging for low-income households, which may struggle to afford the increased costs of keeping their homes cool during hot periods.

The need for enhanced cooling infrastructure, such as more efficient air conditioning systems and heat-resistant building materials, represents a significant investment for urban areas. Retrofitting existing buildings and constructing new ones to be more heat-resilient can be expensive and time-consuming. Additionally, the maintenance of cooling systems and replacing outdated infrastructure add to the economic strain on cities.

The indirect costs of the UHI effect include reduced productivity and economic output. During extreme heat events, workers may experience decreased productivity due to heat stress and discomfort. This can lead to economic losses for businesses and reduced economic activity overall.

Moreover, the healthcare costs associated with treating heat-related illnesses and managing the increased burden on healthcare facilities during heatwaves add to the economic impact. Cities must allocate additional resources to public health measures, emergency response systems, and cooling centers to protect vulnerable populations during extreme heat events.

Objectives and Scope of the Book

The objectives and scope of this book are centered on providing a comprehensive guide to understanding and mitigating the UHI effect through the strategic use of cool pavements.

Purpose of the Guide

This guide's primary purpose is to provide a comprehensive and practical resource for understanding and mitigating the UHI effect through the implementation of cool pavements. As cities worldwide continue to expand and urbanize, the UHI effect poses significant challenges to urban sustainability and resilience. This book aims to equip urban planners, policymakers, engineers, environmental professionals, and students with the knowledge and tools to effectively address these challenges.

By focusing on cool pavements, this guide highlights a tangible and effective solution to reduce urban temperatures and improve the overall quality of life in cities. Cool pavements, with their ability to reflect more sunlight and absorb less heat than traditional materials, offer a practical approach to mitigating the UHI effect. This book will explore the science behind cool pavements, their benefits, design and implementation strategies, and their role in broader urban planning and climate adaptation efforts.

Target Audience

The target audience for this book includes a diverse range of professionals and stakeholders involved in urban development, environmental management, and climate change adaptation. Specifically, the guide is designed for:

- Urban planners and designers seeking to incorporate sustainable and resilient infrastructure into city plans.
- Municipal authorities and policymakers responsible for creating and enforcing regulations to combat urban heat.
- Environmental engineers and architects looking for innovative materials and technologies to reduce urban temperatures.
- Climate change professionals and sustainability advocates interested in practical solutions for urban resilience.

- Students and academics in fields related to urban planning, environmental science, and civil engineering.

This guide, by addressing a broad spectrum of readers, fosters a multidisciplinary approach to UHI mitigation, encouraging collaboration across various sectors and disciplines to achieve sustainable urban development.

Structure and Content Overview

This book's structure is designed to provide a thorough and systematic exploration of the UHI effect and the potential of cool pavements as a mitigation strategy. The content is organized into several key sections, each addressing different aspects of the topic.

1. Introduction: The opening chapter sets the stage by introducing the UHI effect, its significance, and the guide's purpose. It provides an overview of the challenges the UHI effect poses and the potential benefits of cool pavements.
2. Understanding Urban Heat Islands: This section delves into the fundamental concepts of UHIs, exploring their causes, impacts, and global significance. It includes detailed explanations of how UHIs form, their effects on urban environments, and case studies of cities experiencing the UHI effect.
3. Cool Pavements: An Effective Solution: This chapter focuses on cool pavements, explaining what they are, their types, and their benefits. It covers reflective, permeable, and vegetated pavements, highlighting their unique characteristics and advantages.
4. Design and Implementation of Cool Pavements: This section covers the practical aspects of planning, designing, and implementing cool pavements. It includes best practices for site assessment, material selection, construction techniques, and maintenance strategies. Case studies of successful projects provide real-world examples and lessons learned.
5. Integrating Cool Pavements into Urban Planning: This chapter emphasizes the importance of incorporating cool

pavements into broader urban planning and policy frameworks. It discusses relevant policies, regulatory frameworks, and strategies for engaging stakeholders and fostering collaboration.
6. Climate Change Adaptation through Cool Pavements: The role of cool pavements in climate change adaptation is explored in this section. It highlights how cool pavements can enhance urban resilience, their synergies with other climate adaptation measures, and examples of climate-resilient cities.
7. Sustainable Cities and Cool Pavements: This chapter discusses the broader context of sustainable urban development and the contribution of cool pavements to urban sustainability. It covers the integration of cool pavements with green infrastructure and the economic and social benefits of such initiatives.
8. Challenges and Barriers to Implementation: The book addresses the technical, financial, policy, and institutional challenges of implementing cool pavements. It provides strategies for overcoming these barriers and examples of successful change management.
9. Future Trends and Innovations: Emerging technologies, future research directions, and the potential for global adoption of cool pavements are covered in this section. It looks at innovative materials and technologies and discusses the future prospects of cool pavement solutions.
10. Conclusion: The final chapter summarizes the key points discussed throughout the book, provides a call to action for urban planners and policymakers, and presents a vision for the future of cool pavements in sustainable cities.

By the end of this guide, readers will have a comprehensive understanding of the UHI effect and the potential of cool pavements as a practical and effective solution. They will be equipped with the knowledge to design, implement, and advocate for cool pavement projects in their cities, contributing to the creation of more sustainable and climate-resilient urban environments.

Chapter 2: Understanding Urban Heat Islands

UHIs represent a critical environmental challenge for modern cities, as they exacerbate the effects of climate change and pose significant risks to public health, infrastructure, and the natural environment. Human activities and the extensive modification of land surfaces primarily drive the phenomenon where urban areas experience higher temperatures than their rural surroundings. As cities expand and develop, natural landscapes are replaced with buildings, roads, and other structures that absorb and retain heat, forming UHIs. Understanding the mechanisms behind the UHI effect, its causes, and its impacts is essential for developing effective mitigation strategies and fostering sustainable urban development.

This chapter provides a detailed exploration of the UHI effect, beginning with a clear definition and examining its causes. It delves into the historical context of UHI, tracing its recognition and study from the early 19th century to the present day. The chapter also discusses the global significance of UHI, highlighting its far-reaching implications for cities worldwide. Through case studies of major cities experiencing UHIs, readers will gain insights into the diverse manifestations of this phenomenon and the various factors contributing to its intensity.

The chapter will cover the following key areas:

1. Definition and Causes of UHI: A thorough explanation of what constitutes an UHI and the primary factors that lead to its formation. This section will explore how different urban materials, human activities, and city designs contribute to elevated temperatures in urban areas.
2. Historical Context and Development: An overview of the historical recognition and study of the UHI effect, starting from the early observations by Luke Howard in the 19th century to contemporary research. This section will highlight

key milestones in the understanding and documentation of UHIs.
3. Global Significance of UHI: Discuss the importance of addressing the UHI effect globally. This section will cover the impacts of UHI on public health, the environment, and urban economies, emphasizing the need for comprehensive mitigation strategies.
4. Case Studies of UHI in Major Cities: Detailed case studies of cities that have experienced significant UHI effects. These examples will illustrate the diversity of UHI manifestations and the unique challenges different urban areas face. The case studies will also showcase successful mitigation efforts and the lessons learned from these initiatives.

By the end of this chapter, readers will have a comprehensive understanding of the UHI effect, including its causes, historical development, and global significance. This foundational knowledge will set the stage for exploring practical solutions, such as cool pavements, to mitigate the UHI effect and promote sustainable urban environments.

Definition and Causes of UHI

The UHI effect, characterized by significantly higher temperatures in urban areas than their rural surroundings, arises from a combination of human activities and changes in land use.

Explanation of the UHI Phenomenon

The UHI effect is a well-documented phenomenon where urban areas experience significantly higher temperatures than their rural counterparts. This temperature differential is primarily due to the alteration of natural landscapes and the concentration of human activities in urban settings. The UHI effect is most pronounced during the evening and nighttime when urban surfaces that have absorbed heat throughout the day release it slowly, maintaining elevated temperatures in the city compared to the cooler rural areas.

The UHI effect can be observed in small and large urban areas, though it tends to be more severe in densely populated cities with extensive built environments. The temperature difference between urban and rural areas can range from a few degrees to over ten degrees Celsius, depending on various factors such as city size, population density, and local climate conditions. This localized warming affects the comfort and health of urban residents and has broader implications for energy consumption, air quality, and climate resilience.

Factors Contributing to UHI

Several factors contribute to the development and intensity of the UHI effect. Understanding these factors is crucial for developing effective mitigation strategies. The primary contributors include:

1. Surface Materials: Urban areas are characterized by surfaces such as asphalt, concrete, and buildings, which have high thermal mass and low reflectivity. These materials absorb significant solar radiation during the day and release it slowly at night, contributing to higher temperatures. In contrast, natural landscapes, such as forests and grasslands, have higher reflectivity and lower thermal mass, allowing them to remain cooler.
2. Lack of Vegetation: Vegetation plays a critical role in regulating temperatures through the processes of shading and evapotranspiration. Trees and plants provide shade, reducing the amount of heat absorbed by surfaces, and release moisture into the air, which cools the environment through evaporation. Urban areas with sparse vegetation and green spaces lack these cooling effects, leading to higher temperatures.
3. Human Activities: The concentration of human activities in urban areas generates heat. This includes heat emitted from vehicles, industrial processes, and buildings, particularly air conditioning units and heating systems. The cumulative effect of these heat sources contributes to the overall warming of the urban environment.

4. Urban Geometry: The design and layout of cities also play a role in the UHI effect. Urban canyons, created by closely spaced buildings, trap heat, and reduce airflow, preventing heat dissipation and leading to higher temperatures. The density and height of buildings influence the amount of sunlight absorbed and heat removal efficiency through wind circulation.
5. Pollution and Greenhouse Gases: Urban areas often have higher levels of air pollution, which can contribute to the UHI effect. Pollutants such as black carbon and particulate matter absorb sunlight and increase the heat retained in the atmosphere. Additionally, the concentration of greenhouse gases in urban areas can exacerbate warming by trapping heat.
6. Water Bodies and Moisture: The presence or absence of water bodies and moisture in the urban environment affects local temperatures. Natural water bodies like rivers and lakes provide cooling through evaporation. However, urbanization often leads to reducing or altering these natural features, reducing their cooling benefits.

Examples of UHI in Different Climatic Zones

The UHI effect manifests differently across various climatic zones, influenced by local weather patterns, geography, and urban characteristics. Here are some examples of how UHI presents in different climates:

- Temperate Zones: In temperate regions, cities such as New York City and London experience significant UHI effects. These cities have dense urban cores with extensive use of heat-absorbing materials like concrete and asphalt. During summer, the temperature difference between urban and rural areas can be substantial, leading to increased energy demand for cooling and heightened health risks during heatwaves.
- Tropical Zones: Cities in tropical climates, such as Singapore and Mumbai, also face pronounced UHI effects. High humidity and urban heat can create severe discomfort and

health issues. In these regions, the lack of green spaces and the extensive use of air conditioning further contribute to the UHI effect, creating a feedback loop of increased energy use and urban warming.
- Arid Zones: In arid climates, cities like Phoenix and Dubai experience extreme UHI effects due to the scarcity of vegetation and high reliance on air conditioning. The intense solar radiation in these regions exacerbates the heat retention in urban materials, leading to exceptionally high urban temperatures. Strategies such as cool roofs and reflective pavements are particularly important to mitigate the UHI effect in these environments.
- Cold Zones: The UHI effect can be significant even in colder climates. Cities like Moscow and Montreal may experience UHI effects during their short summers. The heat generated from buildings, especially during winter, can also contribute to the UHI effect, albeit to a lesser extent than warmer climates. The UHI effect can influence snowmelt patterns and local climate conditions in these regions.

Impacts of UHI on Urban Environments

The UHI effect significantly impacts urban environments by elevating temperatures, posing public health risks, and leading to various environmental consequences.

Effects on Urban Temperatures

The UHI effect significantly elevates temperatures in urban areas compared to rural ones. This temperature increase is most noticeable during the evening and nighttime when urban surfaces release the heat they have absorbed throughout the day. In some cities, this can lead to temperature differences of up to 10 degrees Celsius or more between urban and rural areas.

These elevated temperatures create a persistent heat load in urban environments, making cities considerably warmer than their rural counterparts. This can be particularly problematic during heatwaves, as the already higher baseline temperatures in cities are exacerbated, leading to extreme heat conditions. This persistent warmth can also affect the daily lives of urban residents, making outdoor activities uncomfortable and sometimes hazardous, especially during the summer months.

Increased urban temperatures also result in higher energy demands. The need for air conditioning and other cooling methods intensifies as temperatures rise. This heightened energy consumption contributes to greater greenhouse gas emissions, creating a feedback loop that amplifies the UHI effect and contributes to global climate change.

Implications for Public Health

The UHI effect has profound implications for public health. Elevated urban temperatures can increase heat-related illnesses, such as heat exhaustion, heat cramps, and heatstroke. Vulnerable populations, including the elderly, children, individuals with pre-existing health conditions, and those without access to air conditioning, are particularly at risk.

During extreme heat events, the health impacts can be severe and widespread. Hospitals and emergency services often see a surge in admissions due to heat-related conditions. In severe cases, prolonged exposure to high temperatures can result in fatalities. The increased incidence of heat-related illnesses significantly burdens healthcare systems, requiring additional resources and interventions to manage and mitigate the impacts.

Higher temperatures also exacerbate air quality issues. The UHI effect contributes to ground-level ozone formation, a major smog component. Increased ozone levels can aggravate respiratory conditions such as asthma and chronic obstructive pulmonary disease (COPD). Poor air quality can also lead to other health issues, including cardiovascular problems and reduced lung function, further impacting the well-being of urban residents.

Moreover, the stress caused by prolonged heat exposure can affect mental health, leading to increased anxiety, stress, and other psychological issues. The discomfort and health risks associated with elevated urban temperatures underscore the importance of implementing effective mitigation strategies to protect public health.

Environmental Consequences

The environmental consequences of the UHI effect extend beyond just elevated temperatures. One of the most significant impacts is the increased demand for energy. As cities become warmer, the reliance on air conditioning and other cooling technologies grows, leading to higher energy consumption. This increased energy use not only results in higher utility bills for residents and businesses but also contributes to greater emissions of greenhouse gases and other pollutants, exacerbating global warming and air pollution.

Higher urban temperatures can also alter local weather patterns. For example, the heat generated in cities can affect wind patterns and humidity levels, potentially leading to changes in precipitation. These alterations can have cascading effects on local ecosystems and water resources, affecting the availability and quality of water for urban and surrounding rural areas.

Urban vegetation and green spaces are also impacted by the UHI effect. Elevated temperatures can stress trees, plants, and other vegetation, reducing their health and resilience. This can lead to decreased urban biodiversity and the loss of green spaces, crucial for providing shade, cooling, and recreational areas for city residents. The loss of vegetation also reduces the natural cooling effect of evapotranspiration, further intensifying the UHI effect.

Water bodies within urban areas, such as rivers, lakes, and reservoirs, can experience thermal pollution due to higher urban temperatures. Warmer water temperatures can negatively impact aquatic ecosystems, affecting fish and other aquatic life. Thermal pollution can also lead to the proliferation of harmful algal blooms, which can degrade water quality and pose health risks to humans and animals.

Furthermore, the UHI effect can contribute to the degradation of urban infrastructure. Prolonged exposure to high temperatures can accelerate the wear and tear on buildings, roads, bridges, and other infrastructure components. Materials such as asphalt and concrete can deteriorate more quickly under extreme heat, increasing city maintenance and repair costs.

Case Studies of UHI in Major Cities

Examining the UHI effect in major cities like New York City, Tokyo, and Paris reveals the diverse challenges and mitigation strategies employed to address elevated urban temperatures and their associated impacts.

UHI in New York City

New York City, one of the world's most densely populated urban areas, experiences a pronounced UHI effect. The city's extensive use of concrete, asphalt, and glass, combined with high-rise buildings, creates an environment where heat is readily absorbed and slowly released, particularly at night. The lack of green spaces and the concentration of human activities further exacerbate the heat retention in the city.

During the summer, New York City can experience temperature differences of up to 7 degrees Celsius (12.6 degrees Fahrenheit) between urban and rural areas. This difference is particularly noticeable during heatwaves, which have become more frequent and intense due to climate change. The elevated temperatures increase the demand for air conditioning, leading to higher energy consumption and elevated greenhouse gas emissions. This creates a feedback loop that further intensifies the UHI effect.

The public health implications in New York City are significant. The city has recorded higher rates of heat-related illnesses and fatalities during heatwaves, particularly among vulnerable populations such as the elderly, children, and low-income residents who may not have access to air conditioning. In response, the city has implemented various mitigation strategies, including promoting green roofs, expanding urban green spaces, and the implementation of reflective pavements and cool roofs.

One notable initiative is the NYC CoolRoofs program, which encourages building owners to coat rooftops with reflective materials to reduce heat absorption. Additionally, the city has invested in increasing tree canopy coverage through the MillionTreesNYC initiative, aiming to plant and care for one million new trees across the five boroughs. These efforts are part of a broader strategy to mitigate the UHI effect and enhance the city's resilience to climate change.

UHI in Tokyo

Tokyo, Japan's capital and one of the largest cities in the world, also faces a significant UHI effect. The city's rapid urbanization, high population density, and extensive use of heat-absorbing materials contribute to the elevated temperatures experienced in its urban core. Tokyo's UHI effect is exacerbated by the limited availability of green spaces and the heat generated by the dense concentration of vehicles and industrial activities.

Tokyo's summer temperatures can be several degrees higher than those in surrounding rural areas, with the UHI effect contributing to increased frequency and severity of heatwaves. The city's high humidity levels further amplify the discomfort and health risks associated with elevated temperatures. The increased demand for air conditioning during the hot months leads to higher energy consumption and greater emissions of greenhouse gases and air pollutants.

To combat the UHI effect, Tokyo has implemented a range of measures. One innovative approach is the widespread adoption of water-retentive pavements designed to reduce surface temperatures through evaporation. These pavements and the use of cool roofs and reflective materials on buildings help mitigate heat absorption and lower ambient temperatures.

Tokyo also emphasizes the importance of urban greenery in combating the UHI effect. The city has increased efforts to create and maintain green spaces, including parks, green roofs, and vertical gardens. Promoting rooftop and wall greening initiatives has been particularly effective in providing cooling benefits and enhancing urban biodiversity.

In addition, Tokyo has introduced policies to improve energy efficiency in buildings and transportation, thereby reducing the overall heat output from these sources. Public awareness campaigns and community engagement are also key components of Tokyo's strategy to mitigate the UHI effect and promote sustainable urban living.

UHI in Paris

Paris, the capital of France, is another major city grappling with the challenges posed by the UHI effect. The city's historic architecture, narrow streets, and dense urban fabric contribute to its susceptibility to higher temperatures. Paris has experienced significant UHI effects, particularly during heatwaves, which have become more frequent and severe in recent years.

The UHI effect in Paris can lead to temperature differences of up to 4 degrees Celsius (7.2 degrees Fahrenheit) between urban and rural areas. These elevated temperatures seriously affect public health, as seen during the 2003 European heatwave, which resulted in thousands of heat-related deaths across France, including many in Paris. The city's elderly population and those living in poorly ventilated buildings were particularly affected.

In response to the UHI effect, Paris has implemented various mitigation strategies. The city has increased its focus on expanding green spaces and enhancing urban vegetation. Initiatives such as creating urban forests, promoting green roofs, and planting trees along streets are integral to the city's approach to reducing urban temperatures.

Paris has also experimented with using cool pavements and reflective materials in its urban infrastructure. These measures help to reduce heat absorption and lower surface temperatures, contributing to a cooler urban environment. Additionally, the city has implemented policies to encourage energy efficiency in buildings and promote public transportation, thereby reducing heat emissions from these sources.

The "Paris Resilience Strategy" outlines the city's comprehensive approach to addressing the UHI effect and adapting to climate change. This strategy includes measures to improve the thermal comfort of public spaces, enhance the energy performance of buildings, and increase the resilience of critical infrastructure to extreme heat events.

In conclusion, New York City, Tokyo, and Paris each illustrate the diverse ways the UHI effect impacts major cities worldwide. These case studies highlight the importance of implementing tailored mitigation strategies that address each urban environment's unique characteristics and challenges. By learning from these examples, other cities can develop effective approaches to combat the UHI effect and promote sustainable, resilient urban living.

Chapter 3: Cool Pavements: An Effective Solution

As urban areas grapple with the challenges posed by the UHI effect, innovative solutions are essential to mitigate rising temperatures and their associated impacts. One promising approach is the implementation of cool pavements. These pavements are designed to reflect more sunlight and absorb less heat than traditional paving materials, reducing surface and ambient temperatures in urban environments.

Cool pavements come in various types, each offering unique benefits and applications. Reflective pavements, permeable pavements, and vegetated pavements are commonly used. These materials not only help mitigate the UHI effect but also provide additional advantages such as improved stormwater management, enhanced urban aesthetics, and increased infrastructure durability.

This chapter delves into cool pavements, exploring their characteristics, benefits, and the science behind their effectiveness. It examines the different types of cool pavements, detailing their specific properties and how they can be integrated into urban design. Furthermore, the chapter highlights the environmental, economic, and social benefits of adopting cool pavements, demonstrating their value as a multifaceted solution to urban heat and sustainability challenges.

This chapter also showcases successful implementations of cool pavements in various cities around the world through a series of case studies. These examples provide practical insights into the planning, design, and execution of cool pavement projects, offering valuable lessons for urban planners, policymakers, and engineers.

By understanding the potential of cool pavements, cities can take proactive steps to combat the UHI effect, improve the quality of life for residents, and enhance urban resilience in the face of climate change. This chapter aims to equip readers with the knowledge and tools needed to advocate for and implement cool pavements as a key component of sustainable urban development.

What are Cool Pavements?

Cool pavements are innovative surface materials designed to mitigate the UHI effect by reflecting more sunlight and absorbing less heat than traditional paving materials.

Definition and Characteristics

Cool pavements are specially designed surfaces that help mitigate the UHI effect by reflecting more sunlight and absorbing less heat than traditional paving materials. Unlike conventional pavements, typically made from materials like asphalt and concrete that retain significant amounts of heat, cool pavements are engineered to have higher solar reflectance (albedo) and lower thermal emittance. This means they can reflect more of the sun's energy and emit absorbed heat more efficiently, leading to cooler surface and ambient temperatures.

Cool pavements come in various forms, including reflective coatings, permeable materials, and vegetated surfaces. Reflective coatings can be applied to existing pavements to increase their albedo. Permeable pavements allow water to pass through, providing cooling through evaporation. Vegetated pavements incorporate grass or other plants, which provide shade and further cooling through evapotranspiration. These characteristics make cool pavements a versatile and effective solution for reducing urban heat and improving the overall thermal comfort of city environments.

Mechanisms of Cooling

Cool pavements reduce surface and ambient temperatures through several mechanisms. One primary mechanism is increased reflectivity. Cool pavements absorb less heat by reflecting a greater portion of incoming solar radiation, directly lowering their surface temperatures. This reduction in surface temperature can significantly decrease the heat radiated into the surrounding air, thus lowering ambient temperatures.

Another key mechanism is enhanced thermal emittance. Cool pavements are designed to release absorbed heat more efficiently than traditional materials. This helps to dissipate heat more quickly, preventing it from accumulating and contributing to higher temperatures.

Permeable and vegetated cool pavements provide additional cooling benefits through evapotranspiration. Permeable pavements allow water to infiltrate and evaporate, cooling the surface and the air above it. Vegetated pavements, with plants integrated into the surface, offer shading and cooling through transpiration, where plants release moisture into the air, reducing temperatures and improving the urban microclimate. These combined mechanisms make cool pavements an effective strategy for mitigating the UHI effect and enhancing urban sustainability.

Types of Cool Pavements

Several types of cool pavements offer unique characteristics and benefits that make them effective in reducing urban temperatures and mitigating the UHI effect.

Reflective Pavements

Reflective pavements, also known as high-albedo pavements, are designed to reduce heat absorption and mitigate the UHI effect by reflecting a significant portion of incoming solar radiation.

Description and Materials

Reflective pavements, or high-albedo pavements, are surfaces designed to reflect a significant portion of incoming solar radiation, thereby reducing the amount of heat absorbed by the pavement and the surrounding environment. These pavements are typically made from materials with high solar reflectance, also known as albedo. Common materials used in reflective pavements include concrete, light-colored aggregates, and specially formulated reflective coatings that can be applied to existing surfaces.

Concrete is a widely used for reflective pavements due to its naturally higher albedo than asphalt. Light-colored aggregates, such as crushed limestone or granite, can be incorporated into the pavement mix to enhance reflectivity. Additionally, reflective coatings can be applied to both new and existing pavements. These coatings are usually composed of materials like titanium dioxide, which reflect sunlight and have self-cleaning properties that maintain the pavement's reflectivity over time.

Reflective pavements can be used in various urban applications, including roads, sidewalks, parking lots, and rooftops. The choice of material and application depends on the specific needs of the area and the intended benefits. By selecting materials with high reflectivity, cities can effectively reduce surface temperatures and improve thermal comfort in urban environments.

Benefits and Limitations

Reflective pavements offer numerous benefits, making them a valuable tool in the fight against the UHI effect. One of the primary benefits is the reduction in surface and ambient temperatures. These pavements absorb less heat by reflecting more sunlight, leading to cooler surface temperatures. This, in turn, helps to lower the surrounding air temperature, which can significantly improve thermal comfort for urban residents and reduce the overall heat burden on the city.

Another key benefit of reflective pavements is their potential to reduce energy consumption. Lower surface temperatures mean that buildings and infrastructure require less cooling, decreasing air conditioning demand. This can lead to substantial energy savings and lower greenhouse gas emissions, contributing to climate change mitigation efforts.

Reflective pavements also improve the durability and lifespan of urban infrastructure. By reducing the thermal stress on pavement materials, reflective pavements can help minimize the wear and tear caused by extreme temperatures. This can result in lower maintenance costs and extended service life for roads, sidewalks, and other paved surfaces.

However, there are also limitations to the use of reflective pavements. One potential drawback is the initial cost of materials and installation, which can be higher than traditional paving options. Additionally, reflective pavements may require regular maintenance to maintain their reflective properties over time. Dirt, debris, and wear can reduce the pavement's albedo, necessitating periodic cleaning or recoating.

Another limitation is the potential for glare, which can be a concern in certain applications, such as roadways and parking lots. The high reflectivity of these pavements can create bright surfaces that may cause discomfort or visibility issues for drivers and pedestrians. Careful consideration of the placement and use of reflective pavements is essential to mitigate these concerns.

Permeable Pavements

Permeable pavements are innovative surfaces designed to allow water to pass through them. They reduce surface runoff and promote natural groundwater recharge while also helping to mitigate the UHI effect through evaporative cooling.

Description and Materials

Permeable pavements, or porous or permeable paving systems, are constructed from materials that allow water to infiltrate through the surface and into the ground below. These pavements are typically made from porous asphalt, pervious concrete, or permeable interlocking concrete pavers (PICP). These materials are designed to create a durable surface with void spaces that enable water to pass through.

Porous asphalt has a reduced fine aggregate content compared to traditional asphalt, creating interconnected void spaces that facilitate water infiltration. Pervious concrete, similarly, is produced with a mixture of coarse aggregates and a special binding agent that creates large pores, allowing water to flow through. Permeable interlocking concrete pavers are individual units made of concrete or other durable materials, spaced apart with joints filled with permeable materials such as gravel or sand, enabling water to permeate through the gaps.

These permeable pavement systems are often installed over a layered sub-base of crushed stone or gravel that provides structural support and further aids in water infiltration and storage. This design manages stormwater and reduces the heat retained on urban surfaces, contributing to cooler local environments.

Benefits and Limitations

Permeable pavements offer numerous benefits that make them valuable to urban infrastructure. One of the primary benefits is improved stormwater management. By allowing water to infiltrate through the surface, permeable pavements reduce surface runoff, decrease the burden on stormwater systems, and mitigate flooding risks. This also helps to recharge groundwater supplies and maintain the natural hydrological cycle.

Another significant benefit is the cooling effect through evaporation. As water infiltrates and moves through the permeable pavement, it evaporates, cooling the surface and the surrounding air. This evaporative cooling process helps lower ambient temperatures, contributing to the mitigation of the UHI effect and improving thermal comfort in urban areas.

Permeable pavements also enhance water quality by filtering pollutants as water passes through the pavement and underlying layers. This natural filtration process removes contaminants such as oils, heavy metals, and sediments, improving water quality entering groundwater and nearby water bodies.

However, there are limitations to using permeable pavements. One potential drawback is the higher initial cost of materials and installation compared to traditional paving options. The specialized materials and installation techniques required for permeable pavements can make them more expensive upfront. Additionally, permeable pavements may require regular maintenance to prevent clogging of the void spaces. Debris, sediment, and organic matter can accumulate and reduce the permeability of the pavement, necessitating periodic cleaning or vacuuming to maintain its effectiveness.

Another limitation is the potential for reduced structural strength compared to conventional pavements. While permeable pavements are designed to handle typical pedestrian and light vehicle traffic, they may not be suitable for heavy traffic areas without additional structural support. Engineers must carefully consider the load-bearing requirements and site-specific conditions when designing and installing permeable pavement systems.

Vegetated Pavements

Vegetated pavements, green pavements or grass pavers integrate plant life into pavement systems to provide cooling, enhance aesthetics, and promote environmental sustainability in urban areas.

Description and Materials

Vegetated pavements combine traditional pavements' structural support with the natural cooling and environmental benefits of vegetation. These systems typically consist of a grid structure made from concrete, plastic, or other durable materials, filled with soil, and planted with grass or other suitable vegetation. The grid provides the strength to support pedestrian and light vehicle traffic while allowing plants to grow within the openings.

The materials used in vegetated pavements are chosen for their durability and compatibility with plant growth. Concrete grids are commonly used for their strength and longevity, but plastic grids are also popular for their lightweight properties and ease of installation. The grid structure can be laid over a prepared sub-base with layers of crushed stone or gravel to ensure proper drainage and support.

The soil used in vegetated pavements must be carefully selected to promote healthy plant growth while providing adequate drainage. The choice of vegetation is also crucial; hardy, low-maintenance grasses or ground cover plants that can withstand foot traffic and occasional vehicular load are typically chosen. These plants play a vital role in providing shade, reducing surface temperatures through evapotranspiration, and enhancing the aesthetic appeal of urban areas.

Benefits and Limitations

Vegetated pavements offer numerous benefits, making them an attractive option for urban environments. One of the primary benefits is their ability to reduce surface temperatures. The vegetation provides shade and cools the air through evapotranspiration, where water is absorbed by plant roots and released into the atmosphere. This cooling effect helps mitigate the UHI effect and improves thermal comfort for city residents.

Another significant benefit is improved stormwater management. Vegetated pavements allow rainwater to infiltrate the ground, reducing surface runoff and the burden on stormwater systems. This natural infiltration process helps to recharge groundwater supplies and prevent flooding. Additionally, the vegetation and soil act as natural filters, removing pollutants from the water before reaching groundwater or nearby water bodies, thus improving water quality.

Vegetated pavements also enhance urban biodiversity by providing insect, bird, and other small wildlife habitats. They contribute to greening urban spaces, making cities more livable and aesthetically pleasing. Integrating green infrastructure into urban planning can also foster community and well-being among residents.

However, there are limitations to the use of vegetated pavements. One potential drawback is the need for regular maintenance to ensure the health and appearance of the vegetation. This includes watering, mowing, and occasional reseeding or plant replacement. The maintenance requirements can be more demanding than traditional pavements, particularly in areas with harsh climates or limited water availability.

Another limitation is the potential for reduced load-bearing capacity compared to conventional pavements. While vegetated pavements are suitable for pedestrian paths, driveways, and light traffic areas, they may not be appropriate for roads or parking lots with heavy traffic. Excessive weight can damage the grid structure and vegetation, necessitating careful consideration of the intended use and load requirements during the design phase.

Benefits of Cool Pavements

Cool pavements offer a range of significant benefits that make them valuable in addressing urban heat and promoting sustainable cities. These benefits include reducing surface temperatures, improving stormwater management, and enhancing urban aesthetics.

Reduction in Surface Temperatures

One of the most critical benefits of cool pavements is their ability to reduce surface temperatures in urban areas. Traditional paving materials like asphalt and concrete absorb and retain significant heat, leading to higher surface and ambient temperatures. Cool pavements, however, are designed with materials that reflect more sunlight and absorb less heat. This increased solar reflectance, or albedo, helps to keep the pavement surfaces cooler.

Reflective pavements, for example, can reduce surface temperatures by several degrees Celsius compared to conventional materials. This reduction in surface temperature can profoundly impact the surrounding environment. Cooler pavements help to lower the ambient air temperature, creating a more comfortable and livable urban space. This is particularly important during heatwaves when the UHI effect can exacerbate the already high temperatures in cities.

Cool pavements' cooling effect also reduces the need for air conditioning in nearby buildings. As surface temperatures decrease, the heat radiated into the surrounding air is reduced, leading to lower indoor temperatures. This can result in significant energy savings, as less energy is required to cool buildings. Reducing energy consumption lowers utility bills for residents and businesses and reduces greenhouse gas emissions, contributing to climate change mitigation efforts.

Moreover, cooler surfaces can enhance the durability of pavement materials. High temperatures can cause thermal expansion and contraction, leading to pavement cracks and other damage. By maintaining lower surface temperatures, cool pavements experience less thermal stress, which can extend their lifespan and reduce maintenance costs. This makes cool pavements a cost-effective solution for urban infrastructure.

Improved Stormwater Management

Another major benefit of cool pavements is their ability to improve stormwater management. Permeable cool pavements allow water to infiltrate the surface and into the ground below. This capability significantly reduces surface runoff, a common problem in urban areas with extensive impervious surfaces.

Traditional pavements can contribute to flooding during heavy rain events by preventing water from infiltrating the ground. The resulting runoff can overwhelm stormwater systems, leading to localized flooding and water pollution. Permeable cool pavements address this issue by allowing rainwater to pass through the pavement and be absorbed into the ground. This natural infiltration process helps to recharge groundwater supplies and reduce the burden on stormwater infrastructure.

In addition to reducing runoff, permeable pavements help filter pollutants from stormwater. As water passes through the pavement and the underlying aggregate layers, contaminants such as oil, heavy metals, and sediments are trapped and filtered out. This filtration process improves the quality of water that eventually reaches groundwater supplies and nearby water bodies, contributing to better environmental health.

Permeable pavements can have broader environmental benefits by reducing runoff and improving water quality. By mitigating flooding and enhancing water quality, these pavements help protect aquatic ecosystems and reduce the negative impacts of urban development on natural water bodies. This makes permeable cool pavements integral to sustainable urban water management practices.

Enhanced Urban Aesthetics

Cool pavements also significantly enhance the aesthetics of urban environments. Traditional pavements often contribute to many cities' harsh, heat-absorbing landscapes. In contrast, cool pavements, with their lighter colors and potential for integrating vegetation, can create more visually appealing and pleasant urban spaces.

Reflective pavements, for example, are available in various colors and finishes that can be aesthetically pleasing while still providing the necessary reflective properties. These pavements can create cooler, more inviting streetscapes, public plazas, and pedestrian pathways. The lighter colors of reflective pavements can brighten urban areas, making them more attractive and enjoyable for residents and visitors.

Vegetated pavements, which incorporate grass or other plants into the pavement structure, offer additional aesthetic benefits. These green pavements can transform otherwise barren urban areas into lush, green spaces. Vegetated pavements can be used in parking lots, driveways, and pedestrian areas to provide a natural, park-like setting. The presence of greenery not only improves the visual appeal of urban spaces but also contributes to the overall well-being of city residents by providing opportunities for recreation and relaxation.

Cool pavements' aesthetic improvements can also enhance the sense of community and place in urban areas. Well-designed, attractive urban spaces can foster social interactions and community activities, contributing to a higher quality of life. Public spaces that are cooler, greener, and more inviting encourage people to spend time outdoors, which can have positive social and health impacts.

In conclusion, cool pavements offer many benefits essential to sustainable urban development. By reducing surface temperatures, improving stormwater management, and enhancing urban aesthetics, cool pavements help to create cooler, more resilient, and more attractive cities. These benefits address the immediate challenges posed by the UHI effect and contribute to broader environmental sustainability goals and improved quality of life for urban residents. As cities continue to grow and face the impacts of climate change, adopting cool pavements will play a crucial role in building resilient and sustainable urban environments.

Chapter 4: Design and Implementation of Cool Pavements

The design and implementation of cool pavements are crucial in leveraging their benefits to mitigate the UHI effect, enhance urban sustainability, and improve the quality of life in cities. This chapter delves into the practical aspects of planning, designing, and constructing cool pavements, providing a comprehensive guide for urban planners, engineers, and policymakers.

As cities worldwide face the dual challenges of rising temperatures and increasing urbanization, the need for innovative and effective solutions becomes paramount. Cool pavements offer a promising approach to addressing these challenges, but their success depends on careful design and thoughtful implementation. This chapter aims to equip readers with the knowledge and tools necessary to incorporate cool pavements effectively into urban infrastructure projects.

We will begin by exploring the planning and design considerations essential for successfully implementing cool pavements. This includes site assessment, appropriate materials selection, and cool pavements' integration into existing urban infrastructure. Understanding each site's specific needs and conditions is critical to selecting the right type of cool pavement and ensuring its effectiveness.

Next, we will explore the materials and technologies used in cool pavements. From reflective coatings to permeable and vegetated pavements, each type of cool pavement offers unique advantages and requires specific construction techniques. We will discuss the properties of these materials, their applications, and the best practices for their installation.

The chapter will also cover the construction and maintenance of cool pavements. Proper construction practices are vital to ensuring cool pavements' long-term performance and durability. We will provide guidelines for installation, including substrate preparation, material application, and quality control. Additionally, we will address the maintenance requirements for cool pavements, emphasizing the importance of regular upkeep to maintain their cooling benefits and structural integrity.

Throughout the chapter, we will highlight real-world examples and case studies that illustrate successful implementations of cool pavements. These examples will provide practical insights and lessons from various projects, helping readers understand the challenges and opportunities associated with cool pavement initiatives.

By the end of this chapter, readers will have a thorough understanding of the design and implementation process for cool pavements. They will be equipped with the knowledge to plan and execute cool pavement projects effectively, contributing to developing cooler, more sustainable urban environments. Whether you are an urban planner, engineer, policymaker, or student, this chapter will serve as a valuable resource in your efforts to combat urban heat and promote climate resilience through innovative pavement solutions.

Planning and Designing Cool Pavements

Effective planning and thoughtful design are crucial for successfully implementing cool pavements, ensuring they maximize benefits in reducing urban heat and enhancing sustainability.

Site Assessment and Selection

The successful implementation of cool pavements begins with thorough site assessment and selection. This crucial first step ensures that the chosen pavement solutions are appropriately tailored to the specific conditions and requirements of the location. Effective site assessment and selection involve several key factors that need to be considered:

1. Climate and Weather Patterns: Understanding the local climate and weather patterns is essential for selecting the most suitable type of cool pavement. Areas with high solar radiation and frequent heatwaves benefit significantly from reflective pavements, which can reduce surface temperatures by reflecting sunlight. Conversely, regions with heavy rainfall and potential flooding may require permeable pavements to manage stormwater effectively.
2. Existing Infrastructure: Evaluating the existing infrastructure is critical to determining how cool pavements can be integrated. This involves assessing the current state of roads, sidewalks, parking lots, and other paved surfaces. Identifying areas with deteriorating pavement or frequently exposed to high temperatures and heavy traffic can help prioritize locations for cool pavement installations.
3. Soil and Subsurface Conditions: The condition of the soil and subsurface layers plays a vital role in the performance of cool pavements, especially permeable types. Conducting geotechnical investigations helps determine the soil's load-bearing capacity, permeability, and stability. This information is crucial for designing a pavement system supporting traffic loads, allowing water infiltration, and preventing subsurface erosion.
4. Land Use and Traffic Patterns: Understanding land use and traffic patterns helps select the appropriate type of cool pavement for different areas. High-traffic roads may require durable reflective pavements, while pedestrian pathways and parking lots could benefit from permeable or vegetated pavements. Analyzing traffic volumes and types of vehicles using the area ensures that the pavement design can withstand the anticipated load without compromising performance.
5. Environmental and Regulatory Considerations: Environmental regulations and local policies influence the selection and implementation of cool pavements. Relevant

regulations, such as stormwater management requirements, heat mitigation policies, and environmental protection guidelines, must be reviewed. Compliance with these regulations ensures legal adherence and aligns the project with broader sustainability goals.
6. Community Needs and Preferences: Engaging with the local community to understand their needs and preferences can enhance the acceptance and success of cool pavement projects. Community input can provide valuable insights into areas of concern, such as heat-related discomfort, flooding issues, and aesthetic preferences. Incorporating community feedback into the planning process fosters a sense of ownership and support for the project.

By thoroughly assessing these factors, urban planners and engineers can make informed decisions about where and how to implement cool pavements most effectively. A comprehensive site assessment ensures that the selected pavement solutions address specific local challenges and maximize the community's benefits.

Design Principles and Considerations

Once the site assessment and selection process is complete, the next step is to focus on the design principles and considerations that will guide the implementation of cool pavements. Effective design is essential to ensure the pavement system's long-term performance, durability, and sustainability. Key design principles and considerations include:

1. Material Selection: Choosing the right materials is fundamental to the success of cool pavements. Reflective pavements, for example, should use materials with high albedo, such as light-colored aggregates, reflective coatings, or specially formulated concrete mixes. Permeable pavements require materials that facilitate water infiltration, such as porous asphalt, pervious concrete, or permeable interlocking concrete pavers (PICP). Vegetated pavements should incorporate durable grid structures filled with soil and

suitable vegetation to withstand traffic and environmental conditions.
2. Structural Design: The structural design of cool pavements must consider the expected load-bearing requirements. The pavement must be designed for high-traffic areas to support heavy vehicles without compromising its reflective or permeable properties. This may involve using thicker layers of high-strength materials, reinforced sub-bases, or geotextiles to enhance stability. For pedestrian pathways and light-traffic areas, the design can focus more on maximizing cooling and permeability while ensuring adequate structural support.
3. Drainage and Water Management: Effective drainage and water management are crucial for performing permeable and vegetated pavements. The design should include appropriate grading and sub-base layers to facilitate water infiltration and prevent pooling or flooding. Incorporating underdrain systems or detention basins can help manage excess water and protect the pavement structure. Additionally, the design should consider integrating green infrastructure elements, such as rain gardens or bioswales, to enhance water management and provide additional environmental benefits.
4. Thermal Performance: Maximizing the thermal performance of cool pavements involves optimizing their reflective and emissive properties. This can be achieved by selecting high solar reflectance and thermal emittance materials. Reflective coatings and surface treatments can enhance the albedo of existing pavements. The design should also consider the orientation and shading of the pavement to minimize direct sunlight exposure and further reduce surface temperatures.
5. Maintenance and Durability: Designing for maintenance and durability ensures that cool pavements retain their performance over time. This involves selecting materials resistant to wear, weathering, and chemical exposure. Maintaining their albedo through regular cleaning and recoating is essential for reflective pavements. Permeable pavements require periodic inspection and cleaning to prevent clogging of the void spaces. Vegetated pavements

need ongoing care, such as watering, mowing, and reseeding, to maintain healthy vegetation and functionality.
6. Aesthetic and Functional Integration: Cool pavements should be designed to blend seamlessly with the surrounding urban environment, enhancing both aesthetics and functionality. Reflective and permeable pavements can be used creatively to create visually appealing streetscapes, plazas, and pathways. Vegetated pavements can contribute to urban greening efforts, providing attractive green spaces that improve urban areas' visual appeal and ecological value. The design should also consider accessibility and safety, ensuring the pavements are functional for all users, including pedestrians, cyclists, and vehicles.
7. Cost Considerations: While the initial cost of cool pavements may be higher than traditional options, it is important to consider the long-term benefits and cost savings. These include reduced energy consumption, lower maintenance costs, extended pavement lifespan, and improved environmental quality. A comprehensive cost-benefit analysis can help justify the investment in cool pavements and highlight their value in promoting sustainable urban development.

Adhering to these design principles and considerations, urban planners and engineers can create effective and sustainable cool pavement solutions. Thoughtful design ensures that cool pavements deliver their intended benefits, such as reduced surface temperatures, improved stormwater management, and enhanced urban aesthetics while meeting the specific needs of the site and the community.

Materials and Technologies

The choice of materials and technologies is pivotal in the design and implementation of cool pavements, as they determine the effectiveness of these surfaces in mitigating urban heat and promoting environmental sustainability.

Overview of Materials Used in Cool Pavements

The materials used in cool pavements are specifically chosen for their ability to reflect more sunlight and absorb less heat than traditional paving materials. These materials help to reduce surface temperatures, mitigate the UHI effect, and enhance urban sustainability. The primary materials used in cool pavements include reflective coatings, permeable materials, and vegetated systems.

Reflective Coatings and Materials

Reflective pavements often incorporate materials with high albedo, which means they have a high capacity to reflect solar radiation. Common materials include light-colored aggregates, specially formulated concrete mixes, and reflective coatings. Light-colored aggregates, such as crushed limestone or granite, can be mixed with concrete to create a pavement surface that reflects more sunlight. Reflective coatings, which can be applied to existing pavements, are typically made from materials like titanium dioxide. These coatings enhance reflectivity and possess self-cleaning properties that help maintain their reflective capabilities over time.

Permeable Pavement Materials

Permeable pavements allow water to infiltrate through the surface, reducing surface runoff and promoting groundwater recharge. The primary materials in permeable pavements include porous asphalt, pervious concrete, and permeable interlocking concrete pavers (PICP). Porous asphalt is made by reducing the amount of fine aggregates, creating void spaces that allow water to pass through. Pervious concrete uses a mix of coarse aggregates and a binding agent, resulting in a porous structure that facilitates water infiltration. PICPs are units made of concrete or other durable materials, with gaps filled with permeable materials like gravel or sand.

Vegetated Pavement Systems

Vegetated pavements, also known as green pavements or grass pavers, integrate plant life into the pavement structure. These systems typically consist of a grid made from concrete, plastic, or other durable materials filled with soil and planted with grass or other suitable vegetation. The grid provides structural support for pedestrian and light vehicle traffic, allowing plants to grow within the openings. Vegetated pavements offer cooling benefits through shading and evapotranspiration, making them an attractive option for urban areas seeking to increase green space.

Innovative Technologies and Applications

Advancements in technology have led to the development of innovative materials and applications for cool pavements, enhancing their effectiveness and expanding their use in urban environments. These technologies leverage scientific research and engineering principles to provide superior cooling, durability, and multifunctionality.

Photocatalytic Coatings

Photocatalytic coatings are an advanced technology used in reflective pavements to enhance their cooling properties and provide additional environmental benefits. These coatings contain materials like titanium dioxide, which, when exposed to sunlight, initiates a photocatalytic reaction that breaks down pollutants in the air, such as nitrogen oxides and volatile organic compounds. This not only helps to reduce surface temperatures but also improves air quality in urban areas.

Phase Change Materials (PCMs)

Phase change materials are innovative substances that can store and release thermal energy during phase transitions, such as melting and solidifying. Integrating PCMs into pavement materials can help regulate surface temperatures by absorbing excess heat during the day and releasing it at night. This temperature-regulating capability can significantly enhance the thermal performance of cool pavements, making them more effective in mitigating the UHI effect.

Geopolymer Concrete

Geopolymer concrete is an emerging technology that offers a sustainable alternative to traditional Portland cement concrete. Made from industrial byproducts like fly ash or slag, geopolymer concrete has a lower carbon footprint and can be designed with high albedo properties for reflective pavements. Its durability and resistance to chemical attack make it suitable for a wide range of urban applications, including roads, sidewalks, and parking lots.

Smart Pavements

Smart pavements incorporate sensors and other technologies to monitor and manage their performance in real time. These sensors can measure temperature, moisture levels, and structural integrity, providing valuable data that can be used to optimize pavement maintenance and enhance cooling effectiveness. Smart pavements can also be integrated with other urban infrastructure, such as smart grids and water management systems, to provide holistic solutions for sustainable urban development.

Cool Roofs and Pavements Integration

Integrating cool pavements with cool roofs can amplify the cooling effects and create a more comprehensive approach to mitigating the UHI effect. Like cool pavements, cool roofs are designed to reflect more sunlight and absorb less heat. By implementing both technologies in urban areas, cities can achieve greater reductions in surface and ambient temperatures, leading to improved thermal comfort and energy savings.

Permeable Reactive Barriers

Permeable reactive barriers are an innovative application for permeable pavements that enhance their water management capabilities. These barriers contain reactive materials that can filter and treat stormwater as it infiltrates the pavement. By removing pollutants such as heavy metals and nutrients, permeable reactive barriers help to improve water quality and protect urban water bodies.

In conclusion, the materials and technologies used in cool pavements are critical to their effectiveness in reducing urban heat and enhancing sustainability. Reflective coatings, permeable materials, and vegetated systems each offer unique benefits, while innovative technologies like photocatalytic coatings, phase change materials, and smart pavements push the boundaries of what cool pavements can achieve. As urban areas continue to seek solutions to combat the UHI effect, these advanced materials and technologies will play a pivotal role in creating cooler, more resilient cities.

Construction and Maintenance

Effective construction and ongoing maintenance are critical for ensuring cool pavements' long-term performance and benefits. This section delves into best practices for constructing cool pavements and outlines essential maintenance requirements and techniques to sustain their functionality and effectiveness.

Best Practices for Construction

The successful implementation of cool pavements begins with meticulous planning and adherence to best construction practices. Ensuring the quality and durability of these pavements requires a series of well-coordinated steps, from site preparation to material application and quality control.

1. Site Preparation: Proper site preparation is fundamental to successfully constructing cool pavements. This involves clearing the area of debris, grading the site to ensure proper

drainage, and compacting the subgrade to provide a stable foundation. For permeable pavements, installing an appropriate sub-base layer of crushed stone or gravel is crucial to facilitate water infiltration and storage. This sub-base should be designed to handle the expected load and environmental conditions.
2. Material Selection and Handling: The right materials are vital for achieving the desired cooling effects and ensuring the pavement's longevity. Reflective pavements should use high-albedo materials, such as light-colored aggregates or reflective coatings, while permeable pavements require materials that support water infiltration, like porous asphalt or pervious concrete. Vegetated pavements should incorporate durable grid structures filled with suitable soil and vegetation. Proper handling and storage of these materials on-site are essential to maintain their integrity and performance characteristics.
3. Mixing and Application: Pairing and applying pavement materials must follow precise guidelines to ensure uniformity and effectiveness. This involves accurately measuring and blending components to achieve the desired properties for concrete and asphalt mixtures. Reflective coatings should be applied evenly using appropriate techniques, such as spraying or rolling, to ensure consistent coverage. The grid structures should be securely installed for vegetated pavements, and the soil should be adequately compacted before planting vegetation.
4. Quality Control: Quality control is crucial at every stage of construction to ensure that the pavement meets design specifications and performs as intended. This includes regular inspections and testing of materials, sub-base layers, and the finished pavement surface. Compaction, permeability, and reflectivity tests should be conducted to verify that the pavement meets the required standards. Any defects or inconsistencies identified during these inspections should be promptly addressed to prevent future issues.
5. Installation Timing: The installation timing can significantly impact the performance of cool pavements. For example, installing permeable pavements during dry weather

conditions helps prevent issues with water infiltration and ensures proper curing. Similarly, reflective coatings should be applied under suitable temperature and humidity conditions to ensure optimal adhesion and performance. Coordinating the installation schedule with favorable weather conditions can enhance the durability and effectiveness of cool pavements.
6. Safety Considerations: Safety should be a top priority during the construction of cool pavements. This involves implementing measures to protect workers, such as providing appropriate personal protective equipment (PPE) and ensuring safe working conditions. Additionally, traffic management plans should be in place to minimize disruptions and hazards for both construction workers and the public. Proper signage, barriers, and communication are essential to maintain safety throughout construction.

Maintenance Requirements and Techniques

Maintaining the performance and benefits of cool pavements requires regular and effective maintenance practices. Addressing maintenance needs promptly and systematically helps extend the lifespan of the pavements and ensures their continued contribution to mitigating the UHI effect:

1. Regular Inspections: Routine inspections are essential for identifying and addressing maintenance needs before they escalate into significant problems. Inspections should focus on checking for surface wear, cracks, and any signs of deterioration. For permeable pavements, inspections should also include assessments of infiltration capacity and potential clogging. Vegetated pavements require inspections to ensure healthy plant growth and identify areas needing reseeding or repair.
2. Cleaning and Debris Removal: Maintaining the reflectivity and permeability of cool pavements involves regular cleaning and debris removal. Reflective pavements should be kept clean to maintain their high albedo, which may involve

periodic washing or sweeping to remove dirt, grime, and organic material. Permeable pavements need regular cleaning to prevent clogging of the void spaces, which can be achieved through vacuuming, pressure washing, or specialized cleaning equipment. Vegetated pavements benefit from removing litter and debris obstructing plant growth and water infiltration.
3. Surface Repairs: Timely repairs are crucial for maintaining cool pavements' structural integrity and performance. Cracks, potholes, and other surface damages should be repaired promptly to prevent further deterioration. For reflective pavements, damaged areas may require the reapplication of reflective coatings or replacement of worn-out materials. Permeable pavements should be repaired with materials that maintain their infiltration properties, and vegetated pavements should be reseeded or replanted as needed to ensure continuous coverage and cooling benefits.
4. Vegetation Management: Healthy vegetation is key to the effectiveness of vegetated pavements. This involves regular watering, mowing, and fertilizing to support plant growth. Weeding is also necessary to prevent unwanted plants from competing with the desired vegetation. In regions with seasonal changes, winterizing practices, such as covering sensitive plants or adjusting watering schedules, can help protect the vegetation and ensure it remains healthy year-round.
5. Monitoring and Data Collection: Implementing a monitoring program can provide valuable data on the performance of cool pavements and inform future maintenance activities. This includes tracking surface temperatures, infiltration rates, and overall pavement condition. Advanced technologies, such as embedded sensors, can facilitate continuous monitoring and provide real-time data for proactive maintenance management. Analyzing this data helps identify trends, assess the effectiveness of maintenance practices, and plan for long-term pavement sustainability.
6. Community Engagement: Engaging the community in the maintenance of cool pavements can enhance their effectiveness and longevity. Educating residents about the

benefits of cool pavements and encouraging their participation in maintenance activities, such as reporting damages or participating in community clean-up events, fosters a sense of ownership and collective responsibility. Involving the community also helps build support for future cool pavement projects and sustainability initiatives.

In conclusion, the construction and maintenance of cool pavements are integral to maximizing their benefits and ensuring long-term performance. Best practices in construction, including site preparation, material selection, and quality control, lay the foundation for durable and effective pavements. Regular maintenance, involving inspections, cleaning, repairs, and vegetation management, sustains the pavements' cooling capabilities and extends their lifespan. By adhering to these principles, urban planners, engineers, and communities can successfully implement cool pavements to mitigate the UHI effect and promote sustainable urban development.

Case Studies of Successful Cool Pavement Projects

Examining successful cool pavement projects in various cities provides valuable insights into their implementation and benefits to urban environments.

Project Examples from Different Cities

Successful cool pavement projects in cities like Los Angeles and Chicago demonstrate the practical applications and benefits of these innovative solutions.

Los Angeles, California

In Los Angeles, a city known for its high summer temperatures and significant UHI effect, implementing cool pavements has shown promising results. The Cool Streets LA initiative, launched in 2017, involves applying reflective coatings to asphalt streets to reduce surface temperatures. These coatings, made from a light-colored, high-albedo material, reflect more sunlight, and absorb less heat. In neighborhoods where this treatment has been applied, surface temperatures have been reduced by up to 10 degrees Fahrenheit (5.5 degrees Celsius). This cooling effect not only enhances thermal comfort for residents but also reduces the need for air conditioning, contributing to lower energy consumption and greenhouse gas emissions.

Chicago, Illinois

Chicago's Green Alley program is another notable example of successful cool pavement implementation. This program focuses on retrofitting city alleys with permeable pavements to manage stormwater and reduce heat. The permeable pavements, made from porous asphalt and pervious concrete, allow rainwater to infiltrate the ground, reducing surface runoff and flooding risks. In addition to their water management benefits, these pavements also help lower ambient temperatures through evaporative cooling. The success of the Green Alley program has led to its expansion, with hundreds of alleys retrofitted across the city, demonstrating the scalability and effectiveness of cool pavements in urban settings.

Lessons Learned and Best Practices

These projects offer important lessons and best practices that can guide future implementations of cool pavements in other urban areas:

- Community Engagement: One key lesson from successful cool pavement projects is the importance of community engagement. In Los Angeles and Chicago, involving residents in the planning and implementation phases has been crucial. Educating the community about the benefits of cool pavements and seeking their input fosters a sense of ownership and support for the projects. Community

involvement also helps identify specific areas of concern and ensures that the pavements meet local needs.
- Comprehensive Planning: Another best practice is comprehensive planning and site assessment. Successful projects start with detailed evaluations of the local climate, existing infrastructure, and soil conditions. This information guides the selection of appropriate cool pavement materials and technologies. In Los Angeles, for example, reflective coatings were chosen for areas with high solar exposure, while in Chicago, permeable pavements were selected to address stormwater management needs.
- Regular Maintenance: Maintenance is critical to the long-term success of cool pavements. Los Angeles and Chicago have implemented regular maintenance programs to ensure the pavements continue performing as intended. In Los Angeles, reflective coatings are periodically reapplied to maintain their high albedo, while in Chicago, permeable pavements are regularly cleaned to prevent clogging and maintain infiltration capacity.
- Monitoring and Evaluation: Continuous monitoring and evaluation are essential for assessing the effectiveness of cool pavement projects. Data collection on surface temperatures, energy usage, and stormwater management helps quantify the benefits and identify areas for improvement. In Chicago, the city uses sensors to monitor the performance of permeable pavements, providing valuable feedback for future projects.

In conclusion, the case studies from Los Angeles and Chicago illustrate the potential of cool pavements to mitigate the UHI effect and enhance urban sustainability. The lessons learned from these projects highlight the importance of community engagement, comprehensive planning, regular maintenance, and ongoing monitoring. Following these best practices, cities can successfully implement cool pavements and create more resilient urban environments.

Chapter 5: Integrating Cool Pavements into Urban Planning

Integrating cool pavements into urban planning is essential for maximizing their benefits and ensuring they contribute effectively to sustainable urban development. As cities continue to grow and face the challenges of climate change and urbanization, innovative solutions like cool pavements are crucial in mitigating the UHI effect, improving public health, and enhancing the overall quality of urban life.

This chapter explores the multifaceted role of cool pavements in sustainable urban design, highlighting how they can be seamlessly integrated into existing urban planning principles and contribute to broader urban sustainability goals. Urban planners, policymakers, and stakeholders can better advocate for and implement cool pavement projects by understanding their potential impact.

The chapter begins by examining the role of cool pavements in sustainable urban design, focusing on their integration with urban planning principles and their contribution to urban sustainability goals. This section delves into how cool pavements can be incorporated into urban infrastructure projects, zoning regulations, and land-use planning to create cooler, more livable cities.

Next, the chapter addresses the policy and regulatory frameworks that support the implementation of cool pavements. It discusses relevant policies and regulations at various levels of government and strategies for effectively implementing these policies to promote the widespread adoption of cool pavements. Understanding these frameworks is crucial for navigating the regulatory landscape and securing approvals and funding for cool pavement projects.

Collaboration with stakeholders is another critical aspect of integrating cool pavements into urban planning. This chapter explores the importance of engaging communities, local governments, and private sector partners in the planning and implementation. By fostering strong partnerships and encouraging stakeholder involvement, cities can ensure the success and sustainability of cool pavement initiatives.

Throughout the chapter, real-world examples and case studies illustrate successful integration of cool pavements into urban planning. These examples provide practical insights and best practices that can guide future projects and help cities achieve their sustainability and climate resilience objectives.

By the end of this chapter, readers will have a comprehensive understanding of how to integrate cool pavements into urban planning effectively. They will have the knowledge and tools to advocate for cool pavements, navigate policy and regulatory frameworks, and engage stakeholders to create more sustainable urban environments.

Role of Cool Pavements in Sustainable Urban Design

Cool pavements' pivotal role in sustainable urban design is that they provide an innovative solution to mitigate the UHI effect and contribute to creating more livable and resilient cities.

Integration with Urban Planning Principles

Integrating cool pavements with urban planning principles is essential for creating cohesive, sustainable, and climate-resilient cities. Urban planners must consider various factors and strategies to seamlessly incorporate cool pavements into the broader urban fabric. This involves aligning cool pavement initiatives with existing urban planning principles and ensuring they complement other sustainability efforts.

Enhancing Land Use Efficiency

Urban planning principles emphasize the efficient use of land to accommodate growing populations while minimizing environmental impacts. Cool pavements can play a significant role in this context by being integrated into multifunctional spaces. For instance, parking lots, sidewalks, and public plazas can be designed with cool pavements to reduce heat absorption and improve thermal comfort. By incorporating cool pavements into these spaces, cities can enhance land use efficiency while addressing the UHI effect.

Promoting Green Infrastructure

Green infrastructure is a cornerstone of sustainable urban planning, aiming to create interconnected networks of natural and semi-natural areas that provide ecological, economic, and social benefits. Cool pavements, particularly permeable and vegetated types, can be integrated into green infrastructure networks. Permeable pavements allow for natural water infiltration, reducing stormwater runoff and improving groundwater recharge. Vegetated pavements contribute to urban greenery, providing habitats for wildlife and enhancing biodiversity. By incorporating cool pavements into green infrastructure plans, cities can create synergies that enhance environmental quality and resilience.

Supporting Climate Adaptation and Mitigation

Urban planning must address climate change by incorporating strategies for both adaptation and mitigation. Cool pavements are an effective tool for climate adaptation, as they reduce surface and ambient temperatures, making urban areas more resilient to heatwaves. They also contribute to climate mitigation by lowering energy consumption for cooling buildings and reducing greenhouse gas emissions. Urban planners can integrate cool pavements into climate action plans and resilience strategies to address current and future climate challenges. This alignment ensures that cool pavements are part of a comprehensive approach to building climate-resilient cities.

Enhancing Public Health and Comfort

A key principle of urban planning is to enhance public health and comfort for residents. Cool pavements contribute to this goal by reducing urban temperatures and improving air quality. Lower surface temperatures can decrease the incidence of heat-related illnesses and deaths, particularly during extreme heat events. Additionally, cool pavements can help reduce the formation of ground-level ozone, a harmful air pollutant that exacerbates respiratory conditions. By prioritizing cool pavements in areas with high pedestrian traffic, such as parks, playgrounds, and urban centers, planners can create healthier and more comfortable environments for city dwellers.

Integrating with Transportation Planning

Transportation planning is integral to urban development, focusing on creating efficient, safe, and sustainable mobility systems. Cool pavements, such as roads, bike lanes, and pedestrian pathways, can be integrated into transportation infrastructure. Reflective pavements can be used on roadways to reduce heat absorption and improve safety by minimizing glare. Permeable pavements can be applied in parking lots and bike lanes to manage stormwater and reduce flooding risks. By aligning cool pavement initiatives with transportation planning, cities can enhance the sustainability and functionality of their mobility networks.

Encouraging Community Involvement

Community involvement is a fundamental principle of urban planning, ensuring that development projects reflect residents' needs and preferences. Engaging communities in planning and implementing cool pavement projects can foster a sense of ownership and support. Planners can organize public consultations, workshops, and informational campaigns to educate residents about the benefits of cool pavements and gather their input. This participatory approach helps ensure that cool pavement initiatives are well-received and meet the specific needs of local communities.

Leveraging Policy and Funding Mechanisms

Integrating cool pavements into urban planning requires leveraging policy and funding mechanisms. Urban planners can advocate for policies that support using cool pavements, such as zoning regulations, building codes, and incentives for green infrastructure. Securing funding from public and private sources, including grants, subsidies, and public-private partnerships, is crucial for successfully implementing cool pavement projects. Cities can ensure sustainability and scalability by aligning cool pavement initiatives with policy and funding frameworks.

Contribution to Urban Sustainability Goals

Cool pavements contribute significantly to urban sustainability goals by addressing critical environmental, economic, and social challenges. Their implementation supports creating resilient, energy-efficient, and livable cities, aligning with broader sustainability frameworks and climate action plans.

Environmental Benefits

One of the primary contributions of cool pavements to urban sustainability is their ability to mitigate the UHI effect. Cool pavements reduce surface and ambient temperatures in urban areas by reflecting more sunlight and absorbing less heat. This cooling effect helps lower energy consumption for air conditioning in nearby buildings, reducing greenhouse gas emissions. Lower urban temperatures also diminish the formation of ground-level ozone, a harmful air pollutant that impacts respiratory health. These environmental benefits directly support urban sustainability goals focused on reducing carbon footprints and improving air quality.

Additionally, permeable cool pavements enhance stormwater management by allowing water to infiltrate through the surface and into the ground. This natural infiltration reduces surface runoff, mitigates flooding risks, and replenishes groundwater supplies. By managing stormwater more effectively, permeable pavements help protect water quality and reduce the burden on urban drainage systems. This aligns with sustainability objectives related to water resource management and ecosystem preservation.

Economic Benefits

Cool pavements contribute to urban sustainability by offering long-term economic benefits. The reduction in surface and ambient temperatures decreases the demand for energy, leading to lower utility costs for residents, businesses, and municipalities. Energy savings can be substantial, particularly during peak summer when cooling demands are highest. These savings can be redirected to other sustainability initiatives or fund additional cool pavement projects.

Moreover, the use of cool pavements can extend the lifespan of urban infrastructure. By reducing thermal stress on pavement materials, cool pavements experience less cracking and deterioration. This leads to lower maintenance and repair costs, providing economic benefits to municipalities and property owners. The durability of cool pavements supports urban sustainability goals by promoting the efficient use of resources and reducing the environmental impact of frequent pavement replacements.

Cool pavements also create opportunities for green jobs and economic development. The design, manufacturing, and installation of cool pavements generate employment opportunities in the construction and environmental sectors. These jobs contribute to the local economy and support the growth of sustainable industries. Cool pavements align with sustainability goals to create a sustainable and inclusive economy by fostering economic resilience and promoting green technologies.

Social Benefits

The social benefits of cool pavements are equally important in contributing to urban sustainability. By reducing urban temperatures, cool pavements improve thermal comfort for residents and visitors, making outdoor spaces more enjoyable and accessible. This is particularly beneficial in densely populated cities where public spaces and walkways are crucial in community life. Enhanced thermal comfort can encourage outdoor activities, promote social interactions, and improve overall quality of life.

Cool pavements also contribute to public health by reducing heat-related illnesses and fatalities. Lower surface temperatures decrease the risk of heat stress, heat exhaustion, and heatstroke, particularly among vulnerable populations such as the elderly, children, and those with pre-existing health conditions. Improved air quality from reduced ground-level ozone formation further supports public health outcomes. By addressing these health risks, cool pavements align with urban sustainability goals of enhancing community well-being and resilience.

Furthermore, the aesthetic improvements provided by cool pavements can enhance the visual appeal of urban environments. Reflective pavements, permeable pavements with vegetation, and other aesthetically pleasing designs contribute to beautifying public spaces. Attractive urban landscapes foster a sense of pride and belonging among residents, supporting social cohesion and community engagement.

Policy and Regulatory Frameworks

Policy and regulatory frameworks are essential for supporting the widespread adoption and implementation of cool pavements and ensuring they effectively contribute to urban sustainability and climate resilience goals.

Relevant Policies and Regulations

Cool pavements require a supportive policy and regulatory framework encouraging their adoption and integration into urban planning. Several key policies and regulations at local, regional, and national levels can facilitate the use of cool pavements and ensure their benefits are maximized.

Local Policies and Regulations

At the local level, municipalities play a critical role in promoting cool pavements through zoning ordinances, building codes, and urban planning guidelines. Cities can introduce zoning regulations that mandate or incentivize using cool pavements in new developments and major renovation projects. For instance, requirements for reflective or permeable pavements in high-heat zones or flood-prone areas can be integrated into local zoning laws.

Building codes can also include standards for cool pavement materials and construction techniques. These codes ensure that new infrastructure projects adhere to best practices for reducing surface temperatures and managing stormwater. Municipalities can offer incentives such as tax rebates, grants, or expedited permitting processes for developers who incorporate cool pavements into their projects.

Urban planning guidelines can emphasize the importance of cool pavements in achieving sustainability goals. Comprehensive plans and climate action plans can explicitly include cool pavements as part of broader strategies to mitigate the UHI effect and enhance climate resilience. These plans can set targets for cool pavement implementation and outline specific actions to achieve them.

Regional Policies and Regulations

Regional governments and planning organizations can support cool pavements through coordinated efforts that span multiple municipalities. Regional policies can provide a framework for consistent standards and practices across cities, ensuring a unified approach to climate mitigation and sustainability.

For example, regional planning bodies can develop guidelines that promote using cool pavements in transportation infrastructure projects, such as highways and transit systems. These guidelines can be incorporated into regional transportation plans, ensuring that cool pavements are considered in large-scale infrastructure investments.

Regional agencies can also facilitate knowledge sharing and collaboration among municipalities. By organizing workshops, conferences, and technical assistance programs, they can help local governments understand the benefits of cool pavements and how to implement them effectively. Regional funding programs can provide financial support for pilot projects and research initiatives that explore innovative cool pavement technologies.

National Policies and Regulations

At the national level, government policies and regulations can create an environment for widely adopting cool pavements. National building codes and standards can include requirements for cool pavement materials and construction practices, ensuring consistency and quality across the country.

National governments can also support research and development in cool pavement technologies. Funding studies and pilot projects can advance the understanding of cool pavements' benefits and address technical challenges. Research institutions and universities can play a key role in this effort, collaborating with industry partners to develop and test new materials and applications.

National climate policies can incorporate cool pavements as part of broader strategies to reduce greenhouse gas emissions and enhance climate resilience. For example, national adaptation plans and sustainable development strategies can highlight the role of cool pavements in mitigating the UHI effect and managing stormwater. These policies can set targets for cool pavement adoption and provide a roadmap.

International Policies and Regulations

International frameworks and agreements can also influence the adoption of cool pavements. Global initiatives such as the Paris Agreement and the United Nations Sustainable Development Goals (SDGs) emphasize the importance of sustainable infrastructure and climate resilience. Countries committed to these agreements can integrate cool pavements into national strategies to meet international targets.

International organizations and coalitions can facilitate the exchange of best practices and knowledge on cool pavements. By participating in global networks and partnerships, cities and countries can learn from successful implementations worldwide and adapt strategies to their local contexts.

Strategies for Policy Implementation

Effective policy implementation is critical to the successful adoption and integration of cool pavements in urban environments. Strategies for implementing policies that support cool pavements involve a combination of regulatory measures, incentives, public awareness, and partnerships. These strategies ensure that policies translate into tangible actions and results.

Regulatory Measures

- Updating Building Codes and Standards: One primary strategy for policy implementation is updating building codes and standards to include requirements for cool pavements. This involves setting specific criteria for materials, design, and performance that developers must meet. For instance, building codes can mandate using high-albedo materials or permeable pavements in new construction and major renovation projects. Clear and enforceable standards help ensure consistency and quality in adopting cool pavements.

- Zoning and Land Use Regulations: Municipalities can implement zoning regulations that encourage or require using cool pavements in specific areas. Zoning ordinances can designate high-heat zones where cool pavements are mandatory to mitigate the UHI effect. Similarly, flood-prone areas can be designated for permeable pavements to enhance stormwater management. By integrating cool pavements into land use planning, cities can target areas where their benefits are most needed.

Incentives and Funding

- Financial Incentives: Offering financial incentives is an effective way to encourage the adoption of cool pavements. Municipalities and regional governments can provide tax rebates, grants, or low-interest loans to developers and property owners who implement cool pavement solutions. These incentives can offset the higher initial costs associated with cool pavements and make them more attractive to stakeholders.
- Public-Private Partnerships: Public-private partnerships (PPPs) can leverage the strengths of both sectors to implement cool pavement projects. Governments can collaborate with private companies, nonprofit organizations, and academic institutions to fund, design, and construct cool pavements. PPPs can facilitate access to additional resources, expertise, and innovation, accelerating the implementation process and ensuring the success of cool pavement initiatives.

Public Awareness and Education

- Public Awareness Campaigns: Raising public awareness about the benefits of cool pavements is essential for gaining community support and participation. Public awareness campaigns can include informational materials, workshops, and events that educate residents, businesses, and developers about the positive impacts of cool pavements on urban heat

reduction, stormwater management, and public health. Highlighting successful case studies and pilot projects can illustrate the practical benefits and inspire broader adoption.
- Training and Capacity Building: Providing training and capacity-building programs for urban planners, engineers, and construction professionals is crucial for effective policy implementation. These programs can cover best practices in cool pavement design, construction, and maintenance, ensuring that professionals have the knowledge and skills to implement cool pavement projects successfully. Technical assistance and resources can also support municipalities and developers in navigating regulatory requirements and leveraging incentives.

Monitoring and Evaluation

- Establishing Performance Metrics: Setting clear performance metrics is vital for monitoring the effectiveness of cool pavement policies and projects. Metrics can include surface temperature reductions, energy savings, stormwater infiltration rates, and public health improvements. By establishing baseline data and regular monitoring, cities can assess cool pavements' impact and make data-driven decisions to enhance policy implementation.
- Continuous Improvement: Policy implementation should be an iterative process that incorporates feedback and lessons learned from ongoing projects. Regular evaluations and stakeholder consultations can identify challenges and opportunities for improvement. Adjusting policies and strategies based on evaluation results ensures that cool pavement initiatives remain effective and responsive to changing urban needs and conditions.

Collaboration and Stakeholder Engagement

- Engaging Local Governments and Communities: Successful policy implementation requires collaboration with local governments and community stakeholders. Engaging

municipalities in planning and implementation ensures that policies align with local priorities and capacities. Community involvement through public consultations, participatory planning, and feedback mechanisms fosters a sense of ownership and support for cool pavement projects.
- Leveraging Multi-Stakeholder Platforms: Creating multi-stakeholder platforms that bring together government agencies, private sector partners, academia, and civil society can facilitate knowledge sharing, coordination, and joint action. These platforms can serve as forums for discussing policy challenges, sharing best practices, and developing innovative solutions to promote cool pavements.

Collaboration with Stakeholders

Collaboration with stakeholders is essential for the successful implementation and sustainability of cool pavement projects. It ensures that diverse perspectives and resources are leveraged to maximize the benefits for urban environments.

Engaging Communities and Local Governments

Engaging communities and local governments is crucial for successfully adopting and implementing cool pavement projects. Their involvement ensures that projects meet local needs, gain public support, and integrate seamlessly into the urban fabric.

Importance of Community Engagement

Community engagement is vital for building support and fostering a sense of ownership over cool pavement initiatives. When residents understand the benefits of cool pavements—such as reduced urban heat, improved air quality, and enhanced stormwater management—they are more likely to support and advocate for these projects. Community involvement also helps identify specific local issues, such as areas prone to flooding or heat stress, that cool pavements can address.

Effective community engagement starts with clear and accessible communication. Municipalities can use various channels, such as public meetings, workshops, social media, and informational pamphlets, to educate residents about cool pavements. These communications should highlight the practical benefits, provide details about upcoming projects, and explain how residents can participate in the planning process.

Strategies for Engaging Communities

- Public Consultations: Public consultations are a valuable tool for gathering input and feedback from community members. These meetings allow residents to voice their concerns, ask questions, and provide suggestions. Planners can use this feedback to refine project designs and ensure they address community priorities. Hosting consultations at convenient times and locations, and offering virtual participation options, can increase attendance and inclusivity.
- Educational Campaigns: Educational campaigns can raise awareness about the benefits of cool pavements and how they contribute to urban sustainability. Schools, community centers, and local media can be engaged to disseminate information. Interactive activities, such as demonstrations of cool pavement materials or tours of existing projects, can make the learning experience engaging and memorable.
- Participatory Planning: Involving community members directly in the planning process through participatory planning initiatives can enhance project relevance and acceptance. This might include forming advisory committees, conducting surveys, or organizing design charrettes where residents collaborate with planners and engineers to shape project proposals. Participatory planning ensures that projects reflect the community's values and needs.

Role of Local Governments

Local governments play a pivotal role in implementing cool pavement projects. They can develop policies, allocate funding, and oversee project execution. Effective collaboration between municipal departments, such as public works, planning, and environmental services, is essential for coordinating efforts and streamlining processes.

Strategies for Engaging Local Governments

- Policy Development: Local governments can develop policies that mandate or incentivize the use of cool pavements in new developments and infrastructure projects. By integrating cool pavements into zoning regulations, building codes, and climate action plans, municipalities can create a regulatory environment that supports sustainable urban design. Policies can include requirements for high-albedo materials in certain zones, incentives for green infrastructure, and targets for reducing urban heat.
- Funding and Resources: Allocating funding and resources is critical for the success of cool pavement projects. Local governments can secure grants, allocate budgetary funds, and explore public-private partnerships to finance these initiatives. Providing financial support for pilot projects and research can also help demonstrate the effectiveness of cool pavements and build the case for broader implementation.
- Interdepartmental Collaboration: Collaboration among municipal departments ensures that cool pavement projects are well-coordinated and integrated into broader urban planning efforts. Regular meetings and communication between departments can facilitate sharing information, resources, and expertise. This collaborative approach helps identify synergies between initiatives, such as combining cool pavements with green infrastructure or stormwater management projects.
- Training and Capacity Building: Investing in training and capacity-building programs for municipal staff is essential for effective project implementation. These programs can cover best practices in cool pavement design, construction, and maintenance. By equipping staff with the necessary

knowledge and skills, local governments can ensure high-quality project execution and long-term performance.
- Monitoring and Evaluation: Local governments should establish systems for monitoring and evaluating the performance of cool pavement projects. Collecting data on surface temperatures, energy savings, and stormwater management can help assess the impact of these projects and inform future planning efforts. Regular reporting and transparency in project outcomes can also build public trust and support.

Public-Private Partnerships

Public-private partnerships (PPPs) are instrumental in the successful implementation and sustainability of cool pavement projects. These collaborations leverage the public and private sectors' strengths, resources, and expertise, fostering innovative solutions and ensuring that cool pavement initiatives are effectively integrated into urban environments.

Benefits of Public-Private Partnerships

- Access to Resources: One of the primary advantages of PPPs is the access to a broader pool of financial and technical resources. Public entities often face budget constraints, while private companies can provide additional funding and investment. This financial support is crucial for covering the higher initial costs associated with cool pavement materials and technologies. Moreover, private sector involvement can bring specialized skills, advanced technologies, and innovative approaches that might not be available within public institutions.
- Risk Sharing: PPPs allow for sharing risks associated with cool pavement projects. This includes financial risks, such as cost overruns, revenue shortfalls, and operational risks, such as maintenance and performance issues. By distributing these risks between public and private partners, projects can be more resilient and adaptable to unforeseen challenges. This

risk-sharing mechanism can also make projects more attractive to investors, facilitating the mobilization of capital for large-scale initiatives.
- Enhanced Efficiency: Private companies often bring higher efficiency and expertise in project management, procurement, and execution. Their experience in delivering complex infrastructure projects can help streamline processes, reduce delays, and ensure that projects are completed on time and within budget. This operational efficiency is particularly beneficial for cool pavement projects, where timely implementation is critical to achieving the desired environmental and social benefits.

Strategies for Effective Public-Private Partnerships

- Clear Objectives and Agreements: For PPPs to be successful, it is essential to establish clear objectives and agreements from the outset. This includes defining the roles and responsibilities of each partner, setting measurable goals, and outlining the financial and operational framework. Detailed contracts and agreements should specify the expectations for project delivery, performance standards, maintenance requirements, and revenue sharing. Clear agreements help prevent misunderstandings and ensure that all parties are aligned towards common goals.
- Stakeholder Engagement: Effective stakeholder engagement is critical in PPPs to build trust and support among all involved parties. This includes engaging not only the public and private partners but also the community, local governments, and other relevant stakeholders. Transparent communication and regular updates can help inform all parties about project progress, challenges, and successes. Involving stakeholders in decision-making fosters a sense of ownership and collaboration, enhancing the project's overall success.
- Leveraging Private Sector Expertise: Public entities should leverage the expertise of private partners to enhance project outcomes. This can involve tapping into the private sector's

experience with innovative materials, advanced technologies, and efficient construction methods. Private companies can also provide valuable insights into market trends, customer preferences, and sustainability practices. By harnessing this expertise, public entities can ensure that cool pavement projects are cutting-edge and meet high quality and performance standards.

- Flexible and Adaptive Frameworks: PPPs should be designed with flexibility and adaptability in mind to accommodate changing conditions and emerging opportunities. This includes incorporating mechanisms for regular review and adjustment of project plans, budgets, and timelines. Flexibility allows partners to respond effectively to new challenges, such as regulatory changes, environmental impacts, or technological advancements. Adaptive frameworks also facilitate continuous improvement and innovation, ensuring that projects remain relevant and effective.

- Long-Term Commitment: Successful PPPs require a long-term commitment from public and private partners. This commitment involves ongoing collaboration, investment, and support throughout the project's lifecycle. Long-term partnerships help ensure the sustainability and maintenance of cool pavements, as both parties remain engaged in monitoring performance, addressing issues, and implementing improvements. Long-term commitments also provide stability and confidence for investors and stakeholders, enhancing the project's credibility and success.

Examples of Successful Public-Private Partnerships

- Cool Pavements Program in Los Angeles: In Los Angeles, the Cool Pavements Program has successfully partnered with private companies to implement reflective pavement coatings across various neighborhoods. These partnerships have enabled the city to scale its efforts, leveraging private sector expertise in materials and application techniques. The

program has resulted in significant temperature reductions and increased public awareness of the benefits of cool pavements.
- Green Infrastructure Projects in Chicago: Chicago's Green Alley program is another example of effective PPPs. The city collaborates with private contractors and environmental organizations to retrofit alleys with permeable pavements and green infrastructure. These partnerships have facilitated the deployment of innovative solutions for stormwater management and urban cooling, contributing to the city's sustainability goals.

Chapter 6: Climate Change Adaptation through Cool Pavements

As climate change continues to impact urban environments, cities are increasingly seeking innovative solutions to enhance their resilience and adaptability. One such solution is the implementation of cool pavements, which offer a range of benefits that help mitigate the effects of rising temperatures and extreme weather events. This chapter explores the role of cool pavements in climate change adaptation, highlighting their potential to improve urban resilience and sustainability.

The chapter begins by examining the specific impacts of climate change on urban areas and how cool pavements can play a critical role in addressing these challenges. Urban areas are particularly vulnerable to the effects of climate change, including increased heat, more intense and frequent heatwaves, and severe weather events. Cool pavements, with their ability to reduce surface and ambient temperatures, provide a practical and effective means of adapting to these conditions.

Next, the chapter delves into how cool pavements can enhance urban resilience as part of a broader climate resilience strategy. It discusses how cool pavements contribute to making cities more livable and sustainable, from improving public health to reducing energy consumption and mitigating flood risks. Real-world examples of cities that have successfully integrated cool pavements into their climate adaptation plans will be presented to illustrate these benefits in action.

Finally, the chapter explores the synergies between cool pavements and other climate adaptation measures. It emphasizes the importance of integrated approaches that combine cool pavements with other solutions, such as green infrastructure, sustainable water management, and energy-efficient building practices. By leveraging these synergies, cities can create comprehensive and cohesive strategies for urban climate resilience.

Overall, this chapter aims to provide a detailed understanding of how cool pavements can be vital to climate change adaptation in urban areas. It offers insights into practical implementation strategies, highlights successful case studies, and underscores the importance of integrating cool pavements with other adaptation measures to build resilient and sustainable cities.

Adapting Urban Areas to Climate Change

Adapting urban areas to the challenges posed by climate change is essential for ensuring the long-term resilience and sustainability of cities.

Climate Change Impacts on Cities

Cities worldwide are increasingly feeling the impacts of climate change, which manifest in various ways and pose significant challenges to urban environments. As the global climate continues to warm, urban areas are particularly vulnerable due to their high population densities, extensive infrastructure, and concentrated economic activities. The primary climate change impacts on cities include increased temperatures, more frequent and intense heatwaves, altered precipitation patterns, and heightened risk of extreme weather events.

Increased Temperatures and Heatwaves

One of climate change's most direct and immediate impacts on cities is the rise in average temperatures. Urban areas, already subject to the UHI effect, experience even greater temperature increases than rural areas. The UHI effect results from the extensive use of heat-absorbing materials like asphalt and concrete, combined with limited vegetation and green spaces. These urban areas become even hotter as global temperatures rise, exacerbating the UHI effect.

Heatwaves, characterized by prolonged periods of excessively high temperatures, are becoming more frequent and severe due to climate change. Cities are particularly susceptible to the adverse effects of heatwaves, which can lead to increased incidences of heat-related illnesses and fatalities. Vulnerable populations, such as the elderly, children, and those with pre-existing health conditions, are at heightened risk during extreme heat events. The strain on public health systems and emergency services also intensifies during heatwaves, making it critical for cities to implement effective heat mitigation and adaptation strategies.

Altered Precipitation Patterns

Climate change alters precipitation patterns, leading to changes in rainfall frequency, intensity, and distribution. Cities are experiencing more irregular rainfall patterns, including prolonged droughts and intense, short-duration rainstorms. These changes pose significant challenges for urban water management systems, which are often designed based on historical climate data and may not be equipped to handle the new variability in rainfall.

Droughts can lead to water shortages, affecting water availability for drinking, sanitation, and irrigation. Urban areas may face increased competition for limited water resources, necessitating water conservation measures and developing alternative water sources. Conversely, intense rainstorms can overwhelm urban drainage systems, leading to flooding, property damage, and disruption of essential services. The increased frequency of such events highlights the need for resilient urban infrastructure to manage extreme weather conditions.

Extreme Weather Events

In addition to heatwaves and altered precipitation patterns, cities face a heightened risk of extreme weather events, such as hurricanes, typhoons, and severe storms. These events can cause widespread destruction, including damage to buildings, infrastructure, and transportation networks. The economic impact of extreme weather events is substantial, with cities incurring significant costs for recovery and rebuilding efforts.

The increasing severity of extreme weather events due to climate change necessitates robust disaster preparedness and response strategies. Urban areas must invest in resilient infrastructure, early warning systems, and emergency response plans to minimize the damage and ensure the safety of residents. Building codes and land-use planning regulations must be updated to reflect the increased risks and promote the construction of climate-resilient buildings and infrastructure.

Sea Level Rise

Coastal cities are particularly vulnerable to the impacts of sea level rise, driven by the melting of polar ice caps and the thermal expansion of seawater. Rising sea levels can lead to coastal erosion, flooding, and saltwater intrusion into freshwater systems. These changes threaten the livelihoods of millions living in coastal urban areas and can result in significant economic losses.

Adapting to sea level rise requires comprehensive planning and the implementation of protective measures, such as constructing sea walls, restoring natural coastal barriers like mangroves and wetlands, and developing adaptive land-use policies. Cities must also consider long-term strategies for managed retreat and relocating vulnerable communities to safer areas.

Role of Cool Pavements in Adaptation

Cool pavements play a crucial role in climate change adaptation by addressing several key challenges urban areas face, such as rising temperatures, increased heatwaves, and the need for efficient stormwater management. By integrating cool pavements into urban infrastructure, cities can enhance their resilience to climate impacts and improve the quality of life for their residents.

Mitigating Urban Heat Island Effect

One of the primary roles of cool pavements in climate adaptation is mitigating the UHI effect. Traditional pavements, such as asphalt and concrete, absorb and retain heat, leading to higher surface and ambient temperatures in urban areas. Cool pavements, on the other hand, are designed with higher reflectivity and thermal emittance, which helps reduce the amount of heat absorbed and retained. Cool pavements can significantly lower surface temperatures by reflecting more sunlight and emitting absorbed heat more efficiently.

Lower surface temperatures help to reduce the overall ambient temperature in urban areas, making cities more comfortable during heatwaves. This cooling effect particularly benefits vulnerable populations, such as the elderly, children, and those with pre-existing health conditions, who are more susceptible to heat-related illnesses. By mitigating the UHI effect, cool pavements contribute to public health and safety, reducing the risk of heat stress, heat exhaustion, and heatstroke.

Reducing Energy Consumption

Cool pavements also reduce energy consumption in urban areas. During periods of high temperatures, the demand for air conditioning increases, leading to higher energy usage and strain on the electrical grid. By lowering surface and ambient temperatures, cool pavements can help reduce the need for air conditioning in nearby buildings. This leads to energy savings and decreased greenhouse gas emissions, contributing to climate change mitigation efforts.

Energy savings from reduced air conditioning usage can also translate into financial savings for residents and businesses. Lower energy bills improve the affordability of living and operating in urban areas, making cities more economically resilient. Additionally, reduced energy consumption helps to alleviate the burden on power infrastructure, enhancing the reliability and sustainability of urban energy systems.

Improving Stormwater Management

Another important role of cool pavements in climate adaptation is improving stormwater management. Climate change alters precipitation patterns, leading to more intense and frequent rainstorms, which can overwhelm urban drainage systems and cause flooding. Permeable cool pavements, designed to allow water to infiltrate through the surface, solve this challenge.

Permeable pavements reduce surface runoff by allowing rainwater to seep into the ground, replenishing groundwater supplies and reducing the risk of flooding. This natural infiltration process helps to manage stormwater more effectively, preventing water from accumulating on the surface and causing damage to infrastructure and property. By integrating permeable cool pavements into urban design, cities can enhance their resilience to heavy rainfall events and improve overall water management.

Enhancing Urban Aesthetics and Livability

Cool pavements also contribute to the aesthetic appeal and livability of urban environments. Reflective pavements are available in a variety of colors and finishes that can enhance the visual appeal of streetscapes, public plazas, and pedestrian pathways. Vegetated pavements, which incorporate grass or other plants, add greenery to urban areas, improving the quality of outdoor spaces.

Attractive and comfortable public spaces encourage outdoor activities and social interactions, fostering a sense of community and well-being. Enhanced urban aesthetics can also boost property values and attract businesses and tourism, contributing to the economic vitality of cities.

Supporting Integrated Climate Adaptation Strategies

Finally, cool pavements support integrated climate adaptation strategies by complementing other adaptation measures, such as green infrastructure, urban forestry, and sustainable building practices. By combining cool pavements with these solutions, cities can create comprehensive and cohesive approaches to climate resilience.

For example, cool pavements can be used with green roofs and walls to maximize cooling benefits and reduce energy consumption. Similarly, integrating cool pavements with urban green spaces and tree planting initiatives can enhance shade, further lowering temperatures and improving air quality.

Cool Pavements as a Climate Resilience Strategy

Cool pavements are vital components of climate resilience strategies in urban areas, helping cities adapt to the growing challenges posed by climate change.

Enhancing Urban Resilience

Urban resilience refers to the ability of cities to withstand, adapt to, and recover from adverse events, such as those brought about by climate change. Cool pavements are an innovative solution that significantly enhances urban resilience by addressing key vulnerabilities related to extreme heat, stormwater management, and urban livability. Here's how cool pavements contribute to urban resilience.

Mitigating Extreme Heat

One of the most pressing challenges for urban resilience is heatwaves' increasing frequency and intensity. These events pose serious health risks, strain energy systems, and deteriorate the quality of urban life. Cool pavements mitigate the UHI effect by using materials that reflect more sunlight and absorb less heat than traditional pavements. This reduction in heat absorption lowers surface and ambient temperatures, crucial during heatwaves.

Cool pavements decrease the temperature of urban environments, reducing the incidence of heat-related illnesses and fatalities, especially among vulnerable populations such as the elderly, children, and individuals with pre-existing health conditions. Cooler surfaces also lessen the demand for air conditioning, reducing energy consumption and preventing power outages during peak usage times. This energy efficiency contributes to climate mitigation efforts and enhances the reliability of urban power grids during extreme heat events.

Managing Stormwater and Reducing Flood Risks

Climate change has altered precipitation patterns, leading to more intense and frequent rainstorms that can overwhelm urban drainage systems and cause flooding. Cool pavements, particularly permeable types, are crucial in stormwater management and flood prevention. Permeable pavements allow water to infiltrate through the surface, reducing surface runoff and promoting groundwater recharge.

This capability to effectively absorb and manage stormwater helps cities handle heavy rainfall events without severe flooding. By integrating permeable cool pavements into streets, parking lots, and pedestrian pathways, cities can mitigate the impact of heavy rains, protect infrastructure, and reduce the risk of property damage. This not only enhances urban resilience but also supports sustainable water management practices.

Improving Air Quality

Cool pavements contribute to improved air quality by reducing the formation of ground-level ozone, a harmful air pollutant that exacerbates respiratory conditions. High urban temperatures accelerate the chemical reactions that produce ground-level ozone. By lowering surface and ambient temperatures, cool pavements help decrease this pollutant's concentration, leading to better air quality and public health outcomes.

Enhanced air quality has wide-ranging benefits for urban resilience. It reduces healthcare costs associated with treating respiratory and cardiovascular diseases, improves the overall health and productivity of the population, and makes urban areas more attractive and livable.

Enhancing Urban Livability and Social Cohesion

Cool pavements enhance urban spaces' aesthetic appeal and comfort, making them more enjoyable for residents and visitors. Reflective pavements can be designed in various colors and finishes, contributing to visually pleasing streetscapes and public plazas. Vegetated pavements add greenery to urban environments, creating cooler, shaded areas that encourage outdoor activities and social interactions.

Cool pavements make public spaces more comfortable and attractive, fostering social cohesion and community well-being. People are more likely to use and enjoy outdoor spaces when they are cooler and more pleasant, leading to stronger community bonds and a higher quality of life.

Supporting Integrated Climate Adaptation Strategies

Cool pavements are most effective when integrated into broader climate adaptation strategies. They complement other resilience measures, such as green roofs, urban forestry, and energy-efficient buildings, creating a synergistic approach to urban resilience. For example, combining cool pavements with increased tree planting can enhance shading and cooling effects, reducing urban temperatures, and improving air quality.

Additionally, integrating cool pavements into urban planning and policy frameworks ensures their benefits are maximized and sustained over the long term. This holistic approach helps cities build comprehensive resilience plans that address multiple climate-related challenges simultaneously.

Examples of Climate-Resilient Cities Using Cool Pavements

Several cities worldwide have implemented cool pavements as part of their climate resilience strategies, demonstrating the practical benefits and effectiveness of this innovative solution. These examples illustrate how cool pavements can be integrated into urban planning to address the challenges posed by climate change, such as extreme heat and stormwater management.

Los Angeles, California

Los Angeles is a leading example of a city that has embraced cool pavements to combat the UHI effect and improve urban resilience. The Cool Streets LA initiative, launched in 2017, involves applying reflective coatings to asphalt streets across various neighborhoods. These coatings, made from a light-colored, high-albedo material, reflect more sunlight, and absorb less heat, resulting in cooler surface temperatures.

The project has shown impressive results, with surface temperatures in treated areas reduced by up to 10 degrees Fahrenheit (5.5 degrees Celsius) compared to untreated streets. This significant temperature reduction helps to lower ambient air temperatures, reduce energy consumption for cooling, and improve thermal comfort for residents. By targeting neighborhoods particularly vulnerable to heatwaves, the Cool Streets LA initiative enhances public health and safety, making the city more resilient to extreme heat events.

Chicago, Illinois

Chicago's Green Alley program is another example of integrating cool pavements into urban resilience efforts. Launched in 2006, the program focuses on retrofitting alleys with permeable pavements to manage stormwater and reduce flooding risks. The permeable pavements, made from porous asphalt and pervious concrete, allow rainwater to infiltrate the ground, reducing surface runoff and replenishing groundwater supplies.

In addition to their water management benefits, these pavements help to lower urban temperatures through evaporative cooling. The Green Alley program has transformed over 200 alleys across the city, demonstrating the scalability and effectiveness of cool pavements in enhancing urban resilience. The success of this program has not only improved stormwater management and reduced flooding but also contributed to creating cooler and more livable urban environments.

Phoenix, Arizona

Phoenix, known for its extreme heat, has implemented cool pavements to mitigate the UHI effect and adapt to rising temperatures. The Cool Pavement Pilot Program, launched in 2020, involves applying reflective coatings to residential streets. These coatings are designed to reflect more sunlight and reduce the heat the pavement absorbs.

Initial results from the pilot program indicate that surface temperatures in treated areas are reduced by an average of 10-12 degrees Fahrenheit (5.5-6.6 degrees Celsius) compared to untreated areas. This cooling effect not only improves residents' thermal comfort but also helps lower the demand for air conditioning, reducing energy consumption and greenhouse gas emissions. By focusing on residential neighborhoods, the program aims to protect vulnerable populations from the adverse effects of extreme heat, enhancing the city's resilience to climate change.

Tokyo, Japan

Tokyo has also integrated cool pavements into its urban planning to address the challenges of extreme heat and heavy rainfall. The city has implemented water-retentive pavements, which combine high-albedo materials with water-absorbing properties. These pavements help to reduce surface temperatures by reflecting sunlight and cooling through evaporation.

In addition to their cooling benefits, water-retentive pavements improve stormwater management by absorbing rainwater and reducing surface runoff. This dual functionality is particularly valuable in Tokyo, where the combination of high population density and frequent heavy rainstorms poses significant challenges for urban infrastructure. By integrating cool pavements into public spaces and roadways, Tokyo enhances its resilience to heatwaves and flooding, creating a more sustainable urban environment.

Sydney, Australia

Sydney has undertaken several initiatives to incorporate cool pavements into its urban infrastructure to combat the UHI effect and improve climate resilience. The city has experimented with various cool pavement materials in public plazas, pedestrian pathways, and parking lots, including reflective coatings and permeable surfaces. These projects aim to reduce urban temperatures, improve stormwater management, and enhance the aesthetic appeal of public spaces.

Sydney's cool pavement projects have shown promising results in lowering surface temperatures and reducing energy consumption for cooling. By integrating these solutions into its urban planning and development strategies, Sydney is building a more resilient and sustainable city capable of withstanding the impacts of climate change.

Synergies with Other Climate Adaptation Measures

Integrating cool pavements with other climate adaptation measures creates synergies that enhance urban resilience and sustainability, providing comprehensive solutions to the multifaceted challenges of climate change.

Combining Cool Pavements with Other Solutions

Integrating cool pavements with other climate adaptation measures can create powerful synergies that enhance urban resilience, improve environmental quality, and ensure sustainable urban development. By combining cool pavements with green infrastructure, urban forestry, and sustainable building practices, cities can address multiple climate challenges simultaneously and create more livable urban environments.

Green Infrastructure

Green infrastructure, such as green roofs, bioswales, and rain gardens, can work with cool pavements to provide comprehensive climate adaptation benefits. Green roofs, for example, reduce heat absorption by providing a layer of vegetation that insulates buildings and cools the surrounding air through evapotranspiration. When used alongside cool pavements, green roofs can amplify the cooling effect, reducing urban temperatures and mitigating the UHI effect.

Bioswales and rain gardens enhance stormwater management by capturing and filtering runoff before it reaches the drainage system. These green infrastructure elements can be strategically placed next to permeable cool pavements to maximize water infiltration and reduce flooding risks. By absorbing and slowly releasing stormwater, these systems help prevent overloading of urban drainage networks and improve groundwater recharge.

Urban Forestry

Urban forestry, which involves planting and maintaining trees in urban areas, complements the benefits of cool pavements by providing shade and further cooling the urban environment. Trees can significantly reduce surface temperatures through shading, while their leaves and branches cool the air through evapotranspiration. Urban trees create cooler, more comfortable city microclimates when combined with cool pavements.

Planting trees along streets with cool pavements can create shaded walkways that are more pleasant for pedestrians, encouraging walking and reducing reliance on vehicles. This improves public health by promoting physical activity and helps reduce air pollution and greenhouse gas emissions. Additionally, trees enhance urban aesthetics and provide habitats for wildlife, contributing to biodiversity and ecological health.

Sustainable Building Practices

Sustainable building practices, such as using energy-efficient materials and designs, can be integrated with cool pavements to create more resilient urban environments. Buildings with reflective roofs and walls can work synergistically with cool pavements to reduce overall heat absorption in urban areas. This holistic approach minimizes the UHI effect and lowers cooling energy demand, resulting in significant energy savings and reduced greenhouse gas emissions.

Incorporating cool pavements into sustainable building projects, such as eco-friendly housing developments or green commercial complexes, ensures that these initiatives are comprehensive and address heat management and stormwater control. For instance, a green building project might include reflective pavements in parking areas, permeable walkways, and green roofs, all contributing to a cooler and more sustainable urban environment.

Multi-Functional Public Spaces

Creating multi-functional public spaces that integrate cool pavements with other climate adaptation measures can transform urban areas into hubs of resilience and sustainability. Public parks, plazas, and recreational areas can be designed with cool pavements, shade trees, green roofs, and water features. These elements work together to reduce heat, manage stormwater, and create inviting spaces for residents to gather and enjoy outdoor activities.

For example, a public plaza might feature permeable cool pavements that manage rainwater runoff, shaded seating areas with trees, and green walls that provide additional cooling and aesthetic appeal. Such spaces enhance urban resilience and promote social cohesion and community well-being.

Policy and Planning Integration

Effective integration of cool pavements with other climate adaptation measures requires supportive policies and planning frameworks. Urban planners and policymakers should adopt a holistic approach, incorporating cool pavements into broader sustainability and resilience plans. This involves updating building codes, zoning regulations, and urban design guidelines to encourage using cool pavements in combination with green infrastructure, urban forestry, and sustainable building practices.

Collaboration between government agencies, private sector partners, and community organizations is also essential for successful implementation. By working together, stakeholders can ensure that cool pavement projects are well-coordinated and maximize their synergies with other climate adaptation measures.

Integrated Approaches to Urban Climate Resilience

Adopting integrated approaches to urban climate resilience is crucial for effectively addressing the complex and interconnected challenges posed by climate change. By combining cool pavements with other climate adaptation measures, cities can develop comprehensive strategies that enhance their resilience to extreme heat, manage stormwater more effectively, and create healthier, more sustainable urban environments.

Synergistic Benefits

The synergistic benefits of integrating cool pavements with other climate adaptation measures are significant. Cool pavements, which reduce surface temperatures and mitigate the UHI effect, can be combined with green infrastructure, urban forestry, and sustainable building practices to amplify their impact. These combined efforts can substantially reduce urban temperatures, improve air quality, enhance stormwater management, and increase urban biodiversity.

For example, the cooling effects are multiplied when cool pavements are used alongside green roofs and green walls. Green roofs and walls provide additional insulation and cooling through evapotranspiration, which can lower temperatures even further when combined with the reflective properties of cool pavements. This integrated approach reduces the UHI effect and enhances the energy efficiency of buildings, leading to significant energy savings and reduced greenhouse gas emissions.

Holistic Urban Planning

A holistic approach to urban planning involves integrating cool pavements into broader urban design and development strategies. This means considering the placement and design of cool pavements in conjunction with other green infrastructure elements, such as parks, green corridors, and water features. Urban planners can design multi-functional public spaces incorporating cool pavements, permeable surfaces, trees, and water management systems to create resilient and sustainable urban landscapes.

For instance, an urban park designed with permeable cool pavements can manage stormwater effectively while providing a cool, shaded environment for recreation. Including bioswales and rain gardens can further enhance stormwater management and improve water quality. Such integrated designs not only address climate resilience but also create attractive, functional public spaces that enhance the quality of urban life.

Policy and Regulatory Support

Effective policy and regulatory support are essential for promoting integrated approaches to urban climate resilience. Governments at all levels can develop and implement policies that encourage using cool pavements and other climate adaptation measures. This includes updating building codes and zoning regulations to require or incentivize the incorporation of cool pavements, green roofs, and other sustainable practices in new developments and major renovations.

Additionally, cities can establish climate action plans that outline specific targets and strategies for reducing the UHI effect, improving stormwater management, and enhancing urban green spaces. These plans can provide a roadmap for integrating cool pavements into broader resilience efforts and ensure that projects are aligned with long-term sustainability goals.

Collaborative Partnerships

Collaboration between public and private sectors, as well as community involvement, is crucial for the successful implementation of integrated climate resilience strategies. PPPs can leverage the strengths and resources of both sectors to fund, design, and implement cool pavement projects. Engaging communities through participatory planning ensures that residents' needs and preferences are considered, leading to more effective and widely supported initiatives.

For example, a city might partner with private developers and environmental organizations to create a new residential area incorporating cool pavements, green roofs, and extensive tree planting. Community input during the planning stages can help identify priority areas for cooling and ensure that the project meets local needs.

Monitoring and Adaptation

Continuous monitoring and adaptation are essential components of integrated climate resilience strategies. Cities should establish systems to monitor the performance of cool pavements and other adaptation measures, collecting data on surface temperatures, energy savings, and stormwater management. This data can inform future planning and help identify areas for improvement.

Regular evaluation and adjustment of strategies ensure that urban resilience efforts remain effective and responsive to changing conditions. By continuously learning and adapting, cities can enhance their resilience and ensure that climate adaptation measures provide long-term benefits.

Chapter 7: Sustainable Cities and Cool Pavements

As cities worldwide grapple with the challenges of rapid urbanization and climate change, the pursuit of sustainability has become more critical than ever. Sustainable urban development seeks to balance economic growth, environmental protection, and social well-being to create resilient, livable, and equitable cities. Cool pavements are an innovative solution that aligns with these goals, offering numerous benefits that enhance urban sustainability.

This chapter explores the multifaceted role of cool pavements in promoting sustainable cities. It begins by examining how cool pavements contribute to environmental sustainability by mitigating the UHI effect, reducing energy consumption, and improving air quality. These environmental benefits are crucial for creating urban environments that are resilient to the impacts of climate change.

Next, the chapter delves into integrating cool pavements with green infrastructure initiatives. Combining cool pavements with green roofs, rain gardens, and urban forestry can amplify their benefits and create synergistic effects that enhance overall urban resilience. This section highlights practical examples and case studies demonstrating the successful implementation of these integrated approaches.

The chapter also addresses the economic and social dimensions of cool pavements. It provides a cost-benefit analysis to illustrate the financial viability and long-term savings associated with cool pavement projects. Additionally, it explores the social equity and community benefits, emphasizing how cool pavements can contribute to more inclusive and healthy urban spaces.

By understanding the comprehensive benefits of cool pavements, urban planners, policymakers, and stakeholders can make informed decisions that promote sustainable urban development. This chapter aims to provide a thorough overview of how cool pavements can be leveraged to create cities that are not only environmentally sustainable but also economically viable and socially equitable.

Enhancing Urban Sustainability

Enhancing urban sustainability is a critical goal for modern cities, and cool pavements offer a powerful tool for achieving this objective by providing significant environmental, economic, and social benefits.

Environmental Benefits of Cool Pavements

Cool pavements offer a range of environmental benefits that contribute to urban areas' overall sustainability. By addressing key issues such as the UHI effect, energy consumption, and air quality, cool pavements help create healthier and more resilient cities.

Mitigating the Urban Heat Island Effect

One of the most significant environmental benefits of cool pavements is their ability to mitigate the UHI effect. Traditional pavements, like asphalt and concrete, absorb and retain heat from the sun, leading to higher surface and ambient temperatures in urban areas. On the other hand, cool pavements are designed with materials that reflect more sunlight and absorb less heat. These high-albedo materials significantly reduce surface temperatures.

Cool pavements help reduce urban ambient temperatures by lowering surface temperatures. This cooling effect is especially important during heatwaves, which are becoming more frequent and intense due to climate change. Reducing urban temperatures improves thermal comfort for residents and decreases the demand for air conditioning, leading to energy savings and reduced greenhouse gas emissions.

Reducing Energy Consumption

Cool pavements contribute to significant reductions in energy consumption. By keeping surface and ambient temperatures lower, they reduce the need for air conditioning in nearby buildings. During hot weather, the energy demand for cooling can be substantial, straining the power grid and increasing electricity costs. By mitigating the UHI effect, cool pavements help lower indoor temperatures, decreasing the reliance on air conditioning systems.

This reduction in energy demand translates to lower greenhouse gas emissions, as less energy is required from power plants, which rely on fossil fuels. Consequently, cool pavements play a role in mitigating climate change by reducing the carbon footprint of urban areas. Additionally, the energy savings from decreased air conditioning use can be redirected to other sustainable initiatives, further enhancing the environmental benefits.

Improving Air Quality

Cool pavements also have a positive impact on air quality. High urban temperatures accelerate the formation of ground-level ozone, a major component of smog that poses serious health risks. By lowering urban temperatures, cool pavements help reduce the ozone formation rate, thereby improving air quality. This benefit is particularly important in densely populated urban areas where air pollution is a major concern.

Improved air quality has wide-ranging health benefits, especially for vulnerable populations such as children, the elderly, and individuals with respiratory conditions. By reducing pollutants like ground-level ozone, cool pavements contribute to healthier urban environments, decreasing the incidence of respiratory and cardiovascular diseases associated with poor air quality.

Enhancing Stormwater Management

Permeable cool pavements provide additional environmental benefits through improved stormwater management. Unlike traditional impervious surfaces, permeable pavements allow rainwater to infiltrate the ground, reducing surface runoff and the risk of flooding. This natural infiltration process helps recharge groundwater supplies and filter out pollutants, improving the quality of urban water systems.

Effective stormwater management is crucial for maintaining the health of urban waterways and reducing the burden on municipal drainage systems. By incorporating permeable cool pavements into urban infrastructure, cities can sustainably manage stormwater, preventing water pollution and mitigating flood risks.

Promoting Urban Biodiversity

Cool pavements, particularly vegetated ones, can enhance urban biodiversity by providing habitats for plants and animals. Vegetated pavements incorporate greenery into urban landscapes, creating pockets of natural habitat in otherwise built-up areas. These green spaces support urban wildlife, promote biodiversity, and contribute to the overall ecological health of the city.

Additionally, greenery has been shown to improve mental well-being and quality of life for urban residents. By integrating vegetated cool pavements, cities can create more attractive and livable environments that benefit both people and nature.

Contribution to Sustainable Urban Development

Cool pavements contribute significantly to sustainable urban development by addressing environmental, economic, and social challenges that cities face in the wake of rapid urbanization and climate change. Their integration into urban infrastructure not only mitigates the adverse effects of urbanization but also promotes a healthier, more resilient, and more equitable urban environment.

Environmental Sustainability

One of the primary contributions of cool pavements to sustainable urban development is their ability to mitigate the UHI effect. Traditional pavement materials, such as asphalt and concrete, absorb and retain heat, leading to higher surface temperatures and increased ambient temperatures in urban areas. Cool pavements from high-albedo materials reflect more sunlight and absorb less heat, thereby reducing surface temperatures.

Cool pavements help lower urban temperatures by decreasing the demand for air conditioning, which in turn reduces energy consumption and greenhouse gas emissions. This contributes to the overall goal of mitigating climate change and promoting energy efficiency. Furthermore, the reduced temperatures also lower the formation of ground-level ozone, a harmful air pollutant, thus improving air quality and public health.

In addition to mitigating the UHI effect, permeable cool pavements enhance stormwater management by allowing water to infiltrate the ground. This reduces surface runoff, decreases the risk of flooding, and helps recharge groundwater supplies. Improved stormwater management also prevents the pollution of water bodies, maintaining the health of urban ecosystems.

Economic Sustainability

Cool pavements contribute to economic sustainability by offering long-term cost savings and supporting economic growth. By reducing surface and ambient temperatures, cool pavements lower the energy demand for cooling buildings, lowering utility bills for residents, businesses, and municipalities. The energy savings can be substantial, especially during peak summer when cooling demands are highest.

Moreover, the use of cool pavements can extend the lifespan of urban infrastructure. Traditional pavements are prone to thermal expansion and contraction, leading to cracks and deterioration. With their reduced thermal stress, cool pavements experience less wear and tear, reducing maintenance and repair costs. This long-term durability translates into significant savings for municipalities regarding maintenance budgets and resource allocation.

Implementing cool pavements also creates opportunities for green jobs and stimulates economic activity. The design, production, and installation of cool pavements require skilled labor, generating employment opportunities in the construction and environmental sectors. These jobs contribute to the local economy and support the growth of sustainable industries, fostering economic resilience and innovation.

Social Sustainability

Cool pavements contribute to social sustainability by enhancing the livability and equity of urban environments. By mitigating the UHI effect and reducing urban temperatures, cool pavements improve thermal comfort for residents, particularly during extreme heat events. This is especially important for vulnerable populations, such as the elderly, children, and low-income communities, who are more susceptible to heat-related illnesses and have limited access to air conditioning.

Improved air quality resulting from the reduced formation of ground-level ozone and other pollutants has significant public health benefits. Cool pavements contribute to a healthier population and reduce healthcare costs by lowering respiratory and cardiovascular disease incidence. Cleaner air and cooler temperatures also enhance the quality of life, making urban areas more attractive and comfortable.

Additionally, cool pavements can enhance public spaces' aesthetic appeal and usability. Reflective pavements and vegetated cool pavements create cooler, more inviting environments for outdoor activities, social interactions, and community events. By making public spaces more enjoyable and accessible, cool pavements foster social cohesion and community well-being.

Cool Pavements and Green Infrastructure

Integrating cool pavements with green infrastructure initiatives enhances urban sustainability and resilience by combining the benefits of reduced heat, improved stormwater management, and increased green spaces.

Integration with Green Infrastructure Initiatives

Integrating cool pavements with green infrastructure initiatives creates a synergistic approach that enhances urban areas' environmental, economic, and social sustainability. Green infrastructure refers to a network of natural and semi-natural systems that provide environmental benefits and improve urban resilience. By combining cool pavements with green roofs, bioswales, rain gardens, and urban forestry, cities can amplify the positive impacts of each strategy and create more sustainable and livable environments.

Enhancing Stormwater Management

One of the primary benefits of integrating cool pavements with green infrastructure is improved stormwater management. Permeable cool pavements allow water to infiltrate the surface, reducing surface runoff and preventing flooding. When combined with green infrastructure elements like bioswales and rain gardens, the effectiveness of stormwater management is further enhanced. Bioswales and rain gardens capture and filter runoff, allowing water to percolate into the ground slowly and reducing the load on urban drainage systems. This combination helps mitigate flood risks, recharge groundwater supplies, and maintain the health of urban waterways.

For example, a city park designed with permeable cool pavements and integrated bioswales can manage stormwater more efficiently while creating a pleasant recreational space for residents. The cool pavements reduce heat and provide comfortable surfaces, while the bioswales manage excess water and support vegetation growth. This holistic approach to design ensures that the park remains functional and resilient to heavy rainfall events, protecting the environment and public infrastructure.

Reducing Urban Heat

Green infrastructure elements such as green roofs, green walls, and urban forests can work alongside cool pavements to reduce the UHI effect and lower ambient temperatures. Green roofs and walls provide insulation and cooling through evapotranspiration, which can be significantly enhanced when used with cool pavements. Trees and other vegetation offer shade and further cool the environment through their natural processes.

By integrating these elements, cities can create cooler microclimates that improve thermal comfort for residents and reduce the need for air conditioning. This integrated approach helps lower energy consumption and greenhouse gas emissions, contributing to climate change mitigation efforts. Additionally, green spaces' aesthetic and psychological benefits, combined with the functional advantages of cool pavements, create more attractive and livable urban environments.

Supporting Biodiversity and Ecosystems

Integrating cool pavements with green infrastructure also supports urban biodiversity and ecosystems. Vegetated cool pavements, which incorporate plants into the pavement structure, provide habitats for urban wildlife and contribute to ecological health. These green spaces support pollinators, birds, and other species, promoting biodiversity in urban areas.

Green infrastructure elements like green roofs and urban forests enhance these benefits by creating larger and more connected habitats. This network of green spaces helps sustain urban ecosystems, offering food and shelter for wildlife. The combined effect of cool pavements and green infrastructure ensures that cities mitigate climate impacts and promote ecological resilience and sustainability.

Improving Public Health and Well-Being

The integration of cool pavements with green infrastructure has significant public health benefits. Reduced urban temperatures and improved air quality lower the incidence of heat-related illnesses and respiratory conditions. Green spaces provide opportunities for physical activity, relaxation, and social interaction, contributing to mental and physical well-being.

For instance, an urban plaza featuring reflective cool pavements, shade trees, and green walls offers a comfortable and inviting space for community gatherings and activities. The cooling effects of the pavements and vegetation make the area more usable during hot weather, encouraging residents to spend time outdoors and engage in healthy activities. This integration supports social cohesion and enhances the overall quality of urban life.

Promoting Sustainable Urban Development

Integrating cool pavements with green infrastructure aligns with broader goals of sustainable urban development. These combined strategies address key challenges such as climate resilience, resource efficiency, and social equity. Cities can create resilient and sustainable communities by promoting multifunctional urban spaces that provide environmental, economic, and social benefits.

Urban planners and policymakers can support this integration by developing comprehensive plans and regulations that encourage using cool pavements and green infrastructure. Incentives like grants and tax breaks can motivate developers and property owners to implement these sustainable practices. Public awareness campaigns and community engagement initiatives can also help build support for integrated urban design approaches.

Synergistic Effects

The integration of cool pavements with green infrastructure initiatives generates synergistic effects that enhance urban environments' overall sustainability and resilience. These combined strategies provide multiple benefits greater than the sum of their individual contributions. By leveraging the unique advantages of each approach, cities can create more efficient, livable, and climate-resilient urban spaces.

Enhanced Cooling Effects

One of the most significant synergistic effects of integrating cool pavements with green infrastructure is the enhanced cooling of urban areas. Cool pavements reflect more sunlight and absorb less heat, reducing surface and ambient temperatures. The cooling effect is amplified when combined with green infrastructure elements such as green roofs, green walls, and urban forests. Vegetation provides additional shading and cooling through evapotranspiration, which further lowers temperatures.

For instance, a city block with reflective pavements, tree-lined streets, and green roofs will experience significantly cooler temperatures than an area with only one of these elements. The combined cooling effects reduce the UHI effect, making urban areas more comfortable and reducing the need for air conditioning. This enhances thermal comfort for residents and contributes to lower energy consumption and reduced greenhouse gas emissions.

Improved Stormwater Management

Integrating permeable cool pavements with green infrastructure also enhances stormwater management capabilities. Permeable pavements allow rainwater to infiltrate the ground, reducing surface runoff and decreasing the risk of flooding. When combined with bioswales, rain gardens, and other green infrastructure elements, the effectiveness of stormwater management is significantly improved.

These green infrastructure components capture and filter runoff, allowing it to be absorbed slowly into the ground. This process helps prevent overloading of urban drainage systems and reduces the incidence of flooding. Additionally, the natural filtration provided by these elements improves water quality by removing pollutants before they reach water bodies. This integrated approach to stormwater management ensures that cities are better equipped to handle heavy rainfall events, protecting infrastructure and the environment.

Enhanced Biodiversity and Ecosystem Health

Cool pavements, particularly those incorporating vegetation, can support urban biodiversity by providing habitats for various species. These habitats become part of a larger ecological network when integrated with broader green infrastructure initiatives, such as green roofs and urban forests. This connectivity supports a wider range of wildlife, promoting urban biodiversity and ecosystem health.

For example, vegetated pavements and green roofs can provide nesting sites and foraging opportunities for birds and pollinators. Urban forests and green corridors enhance these benefits by offering more connected habitats. This network of green spaces supports diverse species, contributing to the ecological resilience of urban areas. By promoting biodiversity, cities can create more resilient ecosystems that can better adapt to environmental changes.

Improved Air Quality

The combined effects of cool pavements and green infrastructure also contribute to improved air quality. Cool pavements reduce urban temperatures, which in turn lowers the formation of ground-level ozone, a harmful air pollutant. Vegetation further enhances air quality by filtering airborne pollutants, such as particulate matter and nitrogen dioxide, through their leaves and roots.

For instance, a park with permeable cool pavements and abundant greenery will have better air quality than an area with only traditional pavements. The plants absorb pollutants and release oxygen, creating a healthier urban environment. Improved air quality reduces the incidence of respiratory and cardiovascular diseases, enhancing public health and well-being.

Social and Economic Benefits

Integrating cool pavements with green infrastructure provides significant social and economic benefits. Cooler, greener urban spaces are more attractive and enjoyable for residents, promoting outdoor activities and social interactions. This enhances the quality of life and fosters a sense of community.

Economically, the reduced need for air conditioning leads to lower energy bills for residents and businesses. The long-term durability of cool pavements reduces maintenance costs, and the increased property values associated with green spaces contribute to economic vitality. Additionally, creating and maintaining green infrastructure generate employment opportunities, supporting local economies.

Economic and Social Benefits

The economic and social benefits of cool pavements are substantial. They contribute to the overall sustainability and resilience of urban areas while enhancing the quality of life for residents.

Cost-Benefit Analysis

A comprehensive cost-benefit analysis of cool pavements reveals their long-term economic viability and significant value in enhancing urban sustainability. While the initial installation costs of cool pavements can be higher than traditional materials, the long-term benefits and savings outweigh these upfront expenses. This analysis examines the costs, direct and indirect benefits associated with cool pavements.

Initial Costs

The initial costs of cool pavements can vary depending on the type of materials used and the scale of the project. Reflective pavements, for example, may require specialized coatings or aggregates that are more expensive than conventional asphalt or concrete. Permeable pavements, such as porous asphalt or pervious concrete, also tend to have higher material and installation costs due to the need for a carefully designed sub-base and precise construction techniques.

Vegetated pavements, which incorporate grass or other plants into the pavement structure, involve additional soil preparation, planting, and ongoing maintenance costs. Despite these higher initial costs, the investment in cool pavements is justified by their numerous economic benefits over their lifespan.

Direct Economic Benefits

1. Energy Savings: One of the most significant direct economic benefits of cool pavements is the reduction in energy consumption for cooling buildings. By lowering surface and ambient temperatures, cool pavements reduce the demand for air conditioning, leading to substantial energy savings. This decrease in energy usage translates to lower utility bills for residents, businesses, and municipalities. Over time, these savings can offset the higher initial costs of cool pavement installation.
2. Extended Pavement Lifespan: Cool pavements experience less thermal stress than traditional pavements, resulting in fewer cracks and a longer lifespan. This durability reduces the frequency and cost of maintenance and repairs. Municipalities can save on maintenance budgets and allocate resources more efficiently, enhancing the overall cost-effectiveness of urban infrastructure.
3. Stormwater Management: Permeable cool pavements improve stormwater management by allowing water to infiltrate the ground, reducing surface runoff and decreasing the risk of flooding. This reduces the burden on urban drainage systems and lowers the costs associated with flood damage and stormwater infrastructure maintenance. The

natural filtration process also reduces the need for water treatment, further cutting costs.

Indirect Economic Benefits

1. Increased Property Values: Cool pavements can enhance urban areas' aesthetic appeal and livability, leading to increased property values. Neighborhoods with cooler, more comfortable environments and attractive, well-maintained pavements are desirable to residents and businesses. Higher property values benefit homeowners and can increase the tax base for municipalities, providing additional revenue for public services and infrastructure improvements.
2. Public Health Savings: Cool pavements contribute to better public health outcomes by reducing urban temperatures and improving air quality. Lower rates of heat-related illnesses, respiratory conditions, and cardiovascular diseases reduce healthcare costs for individuals and the public health system. Improved public health also leads to increased productivity and reduced absenteeism, benefiting the economy.
3. Economic Vitality: Cool pavements support the economic vitality of urban areas by creating more pleasant and usable public spaces. Parks, plazas, and streetscapes with cool pavements attract more visitors, encouraging outdoor activities, social interactions, and community events. This increased foot traffic can boost local businesses and stimulate economic activity, fostering a vibrant and dynamic urban environment.
4. Job Creation: Cool pavements' design, production, and installation generate employment opportunities in the construction and environmental sectors. Ongoing maintenance of vegetated pavements and green infrastructure also creates green jobs, supporting local economies and promoting sustainable industry practices.

Long-Term Financial Returns

While the initial costs of cool pavements may be higher, the long-term financial returns are significant. Energy savings, reduced maintenance costs, improved stormwater management, increased property values, public health savings, economic vitality, and job creation all contribute to the overall economic benefits of cool pavements. These returns make cool pavements a sound investment for municipalities seeking to enhance urban sustainability and resilience.

In conclusion, a thorough cost-benefit analysis demonstrates that the economic advantages of cool pavements far outweigh their initial costs. By providing substantial direct and indirect benefits, cool pavements contribute to the long-term financial sustainability of urban areas, making them an essential component of modern, resilient, and sustainable urban infrastructure.

Social Equity and Community Benefits

Cool pavements offer numerous social equity and community benefits that enhance the quality of life for urban residents and contribute to more inclusive, healthy, and vibrant communities. These benefits are particularly significant for vulnerable populations and underserved neighborhoods, where the impacts of climate change and urban heat are often most severe.

Addressing Heat Vulnerability

Urban heat disproportionately affects vulnerable populations, including the elderly, children, low-income families, and individuals with pre-existing health conditions. These groups often have limited access to air conditioning and live in areas with fewer green spaces and higher pollution levels. Cool pavements help mitigate the UHI effect by reducing surface and ambient temperatures, creating cooler and more comfortable environments. This is particularly important during heatwaves, which pose serious health risks to these vulnerable populations.

Cool pavements lower urban temperatures, reducing the incidence of heat-related illnesses and fatalities. This contributes to public health equity by ensuring that all residents, regardless of their socioeconomic status, have access to safer and healthier living conditions. The cooling effects of cool pavements are most impactful in densely populated urban areas where heat exposure is highest, providing critical relief to those most at risk.

Enhancing Public Spaces

Cool pavements improve the usability and appeal of public spaces, making them more inviting and accessible for all community members. Reflective and permeable pavements can be used in parks, playgrounds, plazas, and pedestrian pathways to create cooler and more pleasant environments. This encourages outdoor activities, social interactions, and community gatherings, fostering a sense of community and well-being.

Enhanced public spaces contribute to social cohesion by providing residents with places to connect, relax, and engage in recreational activities. This is particularly valuable in underserved neighborhoods, where access to quality public spaces is often limited. By investing in cool pavements, cities can revitalize these areas, promoting social inclusion and community engagement.

Reducing Health Disparities

Cool pavements reduce health disparities by improving air quality and lowering temperatures. High urban temperatures exacerbate the formation of ground-level ozone and other pollutants, which can lead to respiratory and cardiovascular diseases. Cool pavements help reduce these pollutants by mitigating the UHI effect, improving air quality and health outcomes.

Improved air quality benefits everyone, but it is especially crucial for low-income and minority communities that often bear the brunt of environmental health hazards. Reducing health disparities through better air quality and cooler urban environments ensures that all residents can live healthier lives.

Supporting Economic Opportunities

The implementation and maintenance of cool pavements create economic opportunities within the community. Cool pavements' design, installation, and upkeep generate jobs in the construction and environmental sectors. These green jobs can provide stable employment opportunities for residents, contributing to economic development and poverty reduction.

Additionally, improved aesthetics and functionality in public spaces can stimulate local economies by attracting visitors and encouraging spending at nearby businesses. Enhanced public spaces can boost property values, increase tax revenues, and support local economic growth, benefiting the community.

Promoting Environmental Justice

Cool pavements promote environmental justice by addressing the unequal distribution of environmental benefits and burdens. Historically, low-income and minority communities have been disproportionately affected by environmental hazards, including high heat exposure and poor air quality. By prioritizing the installation of cool pavements in these areas, cities can help rectify these inequalities and ensure that all residents enjoy the benefits of sustainable urban infrastructure.

Environmental justice initiatives that include cool pavements demonstrate a commitment to equitable urban development and the fair distribution of resources. These initiatives can build trust and cooperation between community members and local governments, fostering a sense of shared responsibility and collective action towards sustainability.

Enhancing Climate Resilience

Cool pavements contribute to the overall climate resilience of communities by providing a sustainable and adaptive response to the challenges posed by climate change. Cooler urban environments are more resilient to heatwaves, while improved stormwater management reduces the risk of flooding. These resilience benefits protect all residents, particularly those in vulnerable and underserved areas, from the adverse impacts of climate change.

In conclusion, cool pavements' social equity and community benefits are significant and far-reaching. By reducing heat vulnerability, enhancing public spaces, improving health outcomes, supporting economic opportunities, promoting environmental justice, and enhancing climate resilience, cool pavements contribute to more inclusive, healthy, and vibrant urban communities. Investing in cool pavements is a step towards environmental sustainability and a commitment to social equity and community well-being.

Chapter 8: Challenges and Barriers to Implementation

Implementing cool pavements in urban areas presents various challenges and barriers that must be addressed to realize their full potential for enhancing sustainability and resilience. Despite the clear environmental, economic, and social benefits of cool pavements, various technical, financial, policy, and institutional obstacles can impede their widespread adoption and effective implementation.

This chapter explores these challenges in detail, providing a comprehensive understanding of the hurdles cities and municipalities face when integrating cool pavements into their infrastructure. By identifying these barriers and discussing potential solutions, this chapter aims to equip urban planners, policymakers, and stakeholders with the knowledge to overcome these obstacles and successfully implement cool pavement projects.

We begin by examining the technical and financial challenges associated with cool pavements. This includes addressing technical constraints, such as material performance and climatic suitability, and exploring innovative solutions. Additionally, the chapter delves into funding and financing mechanisms, discussing how cities can secure the necessary resources to support cool pavement initiatives.

Next, we address policy and institutional barriers, highlighting gaps in existing policies and the challenges of implementing new regulations. The importance of institutional coordination and capacity building is emphasized, as effective implementation requires collaboration across multiple government agencies and levels of governance.

Finally, the chapter discusses strategies for overcoming resistance to change, which can be a significant barrier to adopting new technologies and practices. Engaging stakeholders and managing change effectively are crucial for gaining public support and ensuring the long-term success of cool pavement projects. Examples of successful change management from various cities provide practical insights into how these strategies can be applied.

This chapter comprehensively addresses these challenges and barriers and provides a roadmap for overcoming the obstacles to implementing cool pavements, ultimately contributing to the creation of more sustainable, resilient, and livable urban environments.

Technical and Financial Challenges

Technical and financial challenges are among the most significant obstacles to the widespread adoption and implementation of cool pavements in urban areas.

Technical Constraints and Solutions

Implementing cool pavements in urban environments involves overcoming several technical constraints. These challenges can affect cool pavements' performance, durability, and overall feasibility. However, many of these technical issues can be addressed effectively through innovative solutions and careful planning.

Material Performance

One of the primary technical constraints of cool pavements is ensuring that the materials used provide long-term performance and durability. Cool pavements often incorporate reflective coatings or high-albedo materials that can degrade over time due to weathering, traffic wear, and environmental exposure. The degradation of these materials can reduce their effectiveness in reflecting sunlight and mitigating the UHI effect.

To address this, cities can invest in high-quality materials and advanced manufacturing processes that enhance the longevity and resilience of cool pavements. Research and development efforts are crucial in creating more durable reflective coatings and high-albedo materials that withstand harsh urban conditions. Additionally, regular maintenance and reapplication of reflective coatings can help sustain their performance over time.

Climatic Suitability

The performance of cool pavements can vary significantly depending on the local climate. In regions with high precipitation levels, for example, permeable cool pavements may face challenges related to clogging and reduced infiltration capacity. In colder climates, freeze-thaw cycles can impact the structural integrity of certain pavement materials.

Selecting the appropriate type of cool pavement for the specific climatic conditions of the area is essential. In regions prone to heavy rainfall, permeable pavements should be designed with proper sub-base materials and drainage systems to prevent clogging and maintain infiltration capacity. For colder climates, materials resistant to freeze-thaw damage should be used, and additional design considerations, such as proper grading and drainage, should be implemented to manage water effectively and prevent ice formation.

Load-Bearing Capacity

Cool pavements, especially permeable ones, may have lower load-bearing capacities than traditional pavements. This limitation can restrict their use in high-traffic areas, such as major roadways and highways, where the pavement must withstand significant vehicular loads.

To enhance the load-bearing capacity of permeable cool pavements, engineers can incorporate structural supports such as geogrids or reinforced sub-bases. These supports distribute the load more evenly and increase the overall strength of the pavement system. A combination of permeable and non-permeable cool pavements can be used strategically in high-traffic areas, with permeable pavements applied in lower-traffic zones such as parking lots and pedestrian pathways.

Reflective Glare

Another technical issue associated with reflective cool pavements is the potential for glare, which can be a safety concern for drivers and pedestrians. Highly reflective surfaces can create uncomfortable glare, particularly under direct sunlight.

To mitigate reflective glare, the design of cool pavements can incorporate textures or patterns that diffuse sunlight and reduce glare. Alternatively, materials with a moderate level of reflectivity that balance cooling benefits with visual comfort can be selected. Implementing landscaping elements, such as trees and shrubs, alongside reflective pavements can also help minimize glare by providing shading and visual contrast.

Integration with Existing Infrastructure

Integrating cool pavements into existing urban infrastructure can be challenging, particularly in densely built environments where space is limited and disruption to existing services must be minimized.

Careful planning and coordination with urban planners and utility companies are essential for successfully integrating cool pavements. Using modular and flexible pavement designs can facilitate easier installation and maintenance, reducing the impact on existing infrastructure. Pilot projects and phased implementation strategies can also help manage the transition and allow for adjustments based on initial results.

In conclusion, while several technical constraints are associated with implementing cool pavements, innovative solutions and strategic planning can effectively address these challenges. By selecting appropriate materials, designing for specific climatic conditions, enhancing structural support, mitigating glare, and ensuring integration with existing infrastructure, cities can successfully deploy cool pavements to enhance urban sustainability and resilience.

Funding and Financing Mechanisms

Securing adequate funding and developing effective financing mechanisms are crucial for successfully implementing cool pavements in urban areas. The initial costs of cool pavements can be higher than traditional paving materials, which may pose a financial challenge for cities and municipalities. However, various funding and financing strategies can help overcome these barriers, enabling the widespread adoption of cool pavements.

Government Grants and Subsidies

Government grants and subsidies are essential sources of funding for cool pavement projects. Local, state, and federal governments often provide financial support for sustainable infrastructure initiatives that promote environmental and public health benefits. Grants and subsidies can significantly reduce the financial burden on municipalities, making it more feasible to implement cool pavements.

Cities can actively seek grants and subsidies from government agencies dedicated to environmental protection, energy efficiency, and climate resilience. For example, the Environmental Protection Agency (EPA) and the Department of Energy (DOE) in the United States offer various funding programs that support green infrastructure projects. Applying for these grants requires thorough preparation and demonstration of the project's potential benefits, including reducing urban heat, improving air quality, and enhancing stormwater management.

Public-Private Partnerships (PPPs)

Public-private partnerships (PPPs) are an effective financing mechanism that leverages the resources and expertise of both the public and private sectors. By collaborating with private companies, municipalities can access additional funding, innovative technologies, and specialized skills that may not be available within the public sector alone.

Establishing PPPs involves creating agreements where private entities invest in cool pavement projects in exchange for incentives or shared benefits. These partnerships can take various forms, such as design-build contracts, concession agreements, or joint ventures. Private investors may be attracted to PPPs due to the potential for long-term returns on investment, particularly if the cool pavements are part of a broader sustainable development strategy that enhances property values and economic activity.

Tax Incentives and Rebates

Tax incentives and rebates are financial mechanisms that encourage private investment in cool pavements by reducing the overall cost burden for developers and property owners. These incentives can be offered at the local, state, or federal level and can take various forms, such as tax credits, deductions, or rebates for installing cool pavements.

Municipalities can implement tax incentive programs that reward property owners and developers who incorporate cool pavements into their projects. For instance, a city might offer property tax reductions or credits for businesses that install reflective or permeable pavements in their parking lots. These incentives can stimulate private investment in cool pavements, making it financially attractive for stakeholders to adopt sustainable paving solutions.

Green Bonds and Climate Bonds

Green bonds and climate bonds are innovative financing instruments specifically designed to fund projects that benefit the environment and the climate. Issuing green bonds allows municipalities to raise capital for cool pavement projects while demonstrating their commitment to sustainability.

Cities can issue green bonds to finance large-scale cool pavement initiatives. The funds raised from these bonds can be allocated to cover the costs of materials, installation, and maintenance. Investors in green bonds are typically attracted by the environmental impact of their investments and the potential for stable returns. Climate bonds, a subset of green bonds, are specifically geared towards projects contributing to climate resilience, making them an ideal financing tool for cool pavements.

Utility and Energy Efficiency Programs

Utility companies and energy efficiency programs can provide funding and incentives for cool pavement projects. Since cool pavements contribute to energy savings by reducing the need for air conditioning, utility companies have a vested interest in supporting these initiatives.

Municipalities can partner with utility companies to develop incentive programs that fund cool pavement installations. These programs might include rebates for property owners who install cool pavements or direct funding for municipal projects. Additionally, energy efficiency programs, such as those run by the DOE, can provide grants and technical assistance to support the implementation of cool pavements.

Community Financing and Crowdfunding

Community financing and crowdfunding are emerging mechanisms that engage residents and businesses in funding cool pavement projects. By tapping into the community's interest in sustainability, cities can raise funds through small contributions from many supporters.

Municipalities can launch crowdfunding campaigns to finance specific cool pavement projects, such as retrofitting a local park or schoolyard. These campaigns can be promoted through social media and community events, highlighting the project's environmental and social benefits. Community financing initiatives can also involve local businesses sponsoring cool pavement installations in exchange for recognition and marketing opportunities.

In conclusion, various funding and financing mechanisms are available to support the implementation of cool pavements in urban areas. By leveraging government grants, public-private partnerships, tax incentives, green bonds, utility programs, and community financing, cities can overcome financial barriers and invest in sustainable infrastructure that enhances urban resilience and quality of life.

Policy and Institutional Barriers

Policy and institutional barriers pose significant challenges to implementing cool pavements, requiring coordinated efforts and strategic solutions to overcome these obstacles.

Policy Gaps and Implementation Issues

Adopting and implementing cool pavements often encounter several policy-related challenges that can hinder their widespread integration into urban planning. These challenges stem from existing policy gaps, regulatory hurdles, and implementation issues that need to be addressed to facilitate smoother deployment of cool pavement technologies.

Lack of Specific Guidelines

One significant policy gap is the absence of specific guidelines and standards for installing and maintaining cool pavements. Many cities lack clear regulatory frameworks that define the performance criteria, material specifications, and design standards for cool pavements. Without these guidelines, it can be difficult for urban planners and contractors to ensure that cool pavements are installed correctly and perform as expected.

To address this issue, local governments must develop and adopt comprehensive guidelines that specify the requirements for cool pavement projects. These guidelines should include material selection, construction methods, and performance metrics. Training programs for city planners, engineers, and construction workers can help raise awareness and improve understanding of these new standards.

Inadequate Policy Integration

Cool pavements are often not integrated into broader urban sustainability or climate adaptation policies. This lack of integration can result in missed opportunities for funding, support, and synergies with other environmental initiatives. Cool pavement projects may struggle to gain traction and achieve their potential environmental and social benefits without policy support.

Cities should strive to embed cool pavement strategies within existing and new urban development plans and climate action policies. By linking cool pavements to broader goals such as reducing urban heat islands, improving air quality, and enhancing stormwater management, municipalities can prioritize these projects and allocate appropriate resources.

Funding and Financial Incentives

A critical implementation issue is the shortage of dedicated funding and financial incentives for cool pavement projects. Budget constraints and competing priorities often leave little room for innovative infrastructure solutions like cool pavements, which can be more costly than traditional options.

To secure the necessary financial resources, municipalities can explore various funding mechanisms, such as green bonds, grants from environmental agencies, and public-private partnerships. Additionally, offering tax incentives or rebates to property owners and developers who choose cool pavements can encourage private investment in these technologies.

Interdepartmental Coordination

Effective implementation of cool pavements often requires coordination between multiple city departments, including urban planning, environmental protection, public works, and transportation. A lack of coordination can lead to fragmented efforts and inefficiencies that delay or complicate projects.

Establishing a dedicated task force or committee with representatives from all relevant departments can enhance interdepartmental coordination. This task force should oversee the planning, execution, and maintenance of cool pavement projects, ensuring that all efforts are aligned and optimized.

Resistance to Change

Resistance to change among stakeholders, including policymakers, contractors, and the public, can also be a significant barrier. Traditional paving methods are well-understood and trusted, while new technologies like cool pavements may be met with skepticism regarding their cost, effectiveness, and durability.

Education and outreach programs are essential to overcome resistance to change. Municipalities should conduct workshops, seminars, and pilot projects to demonstrate the benefits of cool pavements. Engaging stakeholders early in planning and providing clear, evidence-based information can help build support and alleviate concerns.

By addressing these policy gaps and implementation issues, cities can facilitate cool pavements' broader adoption and success. Developing specific guidelines, integrating policies, securing funding, enhancing coordination, and educating stakeholders are all crucial steps in overcoming the barriers that hinder the implementation of these innovative and beneficial urban infrastructure solutions.

Institutional Coordination and Capacity Building

Effective implementation of cool pavements requires robust institutional coordination and capacity building to ensure that all relevant stakeholders are aligned and equipped with the necessary knowledge and skills. This involves fostering collaboration among various governmental departments, enhancing technical expertise, and building institutional frameworks that support sustainable urban infrastructure.

Enhancing Interdepartmental Coordination

One key challenge in implementing cool pavements is ensuring coordination between different municipal departments, such as urban planning, public works, transportation, and environmental protection. Each of these departments plays a crucial role in planning, designing, implementing, and maintaining cool pavement projects. However, without effective coordination, efforts can become fragmented and inefficient.

Establishing an interdepartmental task force or committee dedicated to cool pavement projects can significantly enhance coordination. This task force should include representatives from all relevant departments and be responsible for overseeing project planning and execution. Regular meetings and clear communication channels can help ensure all departments are aligned and working towards common goals. Additionally, developing integrated planning documents that outline the roles and responsibilities of each department can facilitate smoother project implementation.

Building Technical Expertise

Another critical aspect of capacity building is enhancing municipal staff and contractors' technical expertise in cool pavement projects. The installation and maintenance of cool pavements require specialized knowledge of materials, construction techniques, and performance monitoring, which may not be readily available within existing municipal teams.

Providing targeted training programs and workshops can help build the necessary technical skills. These training sessions can cover various aspects of cool pavements, including material selection, design principles, construction methods, and maintenance practices. Partnering with universities, research institutions, and industry experts can provide additional resources and support for these training programs. Continuous professional development opportunities can also ensure staff stay updated on cool pavement technology's latest advancements and best practices.

Institutionalizing Best Practices

To ensure the long-term success of cool pavement initiatives, it is essential to institutionalize best practices and create frameworks that support sustainable urban infrastructure. This involves developing policies, guidelines, and standard operating procedures integrating cool pavements into regular urban planning and development processes.

Municipalities can develop comprehensive guidelines and manuals that detail the standards for cool pavement projects. These documents should include information on material specifications, design criteria, construction protocols, and maintenance schedules. Incorporating these guidelines into municipal codes and regulations can formalize the adoption of cool pavements and ensure consistent application across projects. Additionally, establishing monitoring and evaluation frameworks can help track the performance of cool pavements and provide data for continuous improvement.

Fostering Collaboration and Partnerships

Building institutional capacity fosters collaboration and partnerships with external stakeholders, including private sector companies, non-profit organizations, and community groups. These partnerships can bring additional resources, expertise, and support to cool pavement projects.

Municipalities can create platforms for collaboration, such as advisory committees or stakeholder forums, where various partners can contribute to the planning and implementation of cool pavements. Engaging local businesses and community organizations in these efforts can also help build public support and ensure that projects are responsive to community needs and priorities.

In conclusion, institutional coordination and capacity building are essential for successfully implementing cool pavements. By enhancing interdepartmental coordination, building technical expertise, institutionalizing best practices, and fostering collaboration, cities can create a supportive environment for sustainable urban infrastructure and ensure the long-term success of cool pavement projects.

Overcoming Resistance to Change

Overcoming resistance to change is crucial for successfully implementing cool pavements, as it requires addressing the concerns and hesitations of various stakeholders through effective engagement and communication strategies.

Strategies for Stakeholder Engagement

Effective stakeholder engagement is essential for overcoming resistance to change and successfully implementing cool pavements. By involving various stakeholders early in the process and addressing their concerns and interests, municipalities can build support and create a collaborative environment for sustainable urban infrastructure projects.

Early and Inclusive Consultation

One of the most effective stakeholder engagement strategies is initiating early and inclusive consultations. Engaging stakeholders from the outset helps identify potential concerns and allows for incorporating their input into the project planning process. This inclusive approach fosters a sense of ownership and commitment among stakeholders, making them more likely to support the initiative.

Municipalities should organize public meetings, workshops, and focus groups to gather input from various stakeholders, including residents, business owners, community organizations, and environmental groups. Providing multiple platforms for engagement, such as online surveys and social media forums, can also ensure broader participation and inclusivity.

Transparent Communication

Transparent communication is key to building trust and reducing resistance to change. Stakeholders need clear and accurate information about cool pavements' benefits, costs, and potential impacts. Addressing misconceptions and providing evidence-based data can help alleviate concerns and build confidence in the project.

Municipalities should develop comprehensive communication plans, including regular updates, informational materials, and public presentations. Creating a dedicated project website or online portal where stakeholders can access relevant information, view progress reports, and submit feedback can enhance transparency and accessibility.

Demonstrating Benefits Through Pilot Projects

Pilot projects can be powerful tools for demonstrating the tangible benefits of cool pavements. By implementing small-scale pilot projects in selected areas, municipalities can showcase the effectiveness of cool pavements in reducing urban heat, improving air quality, and enhancing public spaces. Successful pilot projects can generate positive publicity and build momentum for larger-scale implementation.

Selecting high-visibility locations for pilot projects, such as parks, schoolyards, or busy streets, can maximize their impact. Municipalities should document and share the results of these pilot projects through case studies, video documentaries, and media coverage to highlight the benefits and address potential concerns.

Building Partnerships and Alliances

Forming partnerships and alliances with key stakeholders, such as local businesses, non-profit organizations, academic institutions, and industry experts, can strengthen support for cool pavement projects. These partnerships can provide additional resources, expertise, and advocacy, helping to overcome resistance and facilitate project implementation.

Municipalities should actively seek collaborations with stakeholders who have a vested interest in sustainable urban development. Joint initiatives, co-branded events, and collaborative research projects can help build strong alliances and create a united front supporting cool pavements.

Examples of Successful Change Management

Successful change management in implementing cool pavements can be observed in several cities that have effectively navigated stakeholder resistance and achieved significant progress in sustainable urban infrastructure. These examples highlight the importance of strategic planning, community involvement, and clear communication.

Los Angeles, California

Los Angeles has been a pioneer in the implementation of cool pavements through its Cool Streets LA initiative. The city faced initial resistance from residents concerned about the costs and effectiveness of reflective pavements. To address these concerns, the city launched a series of pilot projects in various neighborhoods, demonstrating the cooling benefits and gathering data to support the initiative.

Los Angeles engaged in extensive community outreach, including public meetings and informational sessions, to educate residents about the benefits of cool pavements. The city also developed partnerships with local universities to conduct research and provide evidence-based data on the effectiveness of the cool pavements. This transparent approach and the visible success of the pilot projects helped build public support and paved the way for broader implementation.

Chicago, Illinois

Chicago's Green Alley program successfully integrated permeable cool pavements into its urban infrastructure, despite initial skepticism from local businesses and residents. Concerns about the durability and maintenance of permeable pavements were prevalent among stakeholders.

The city addressed these concerns by implementing a pilot program in several alleys and closely monitoring the results. They provided clear communication through regular updates and reports, highlighting the improved stormwater management and reduced flooding in these areas. Additionally, Chicago partnered with local businesses and community groups to maintain the permeable pavements, demonstrating a collaborative approach to urban sustainability.

Phoenix, Arizona

Phoenix, known for its extreme heat, implemented the Cool Pavement Pilot Program to combat the UHI effect. The city faced resistance from residents and businesses worried about potential glare and the effectiveness of reflective pavements in such a hot climate.

Phoenix conducted thorough testing to overcome this resistance and selected materials specifically designed to minimize glare while maximizing reflectivity. The city organized community workshops and demonstrations to showcase the benefits of cool pavements and engaged local media to spread positive stories about the pilot program's success. By addressing concerns through targeted solutions and clear communication, Phoenix gained public support and expanded the program.

Sydney, Australia

Sydney's implementation of cool pavements in public parks and pathways initially encountered resistance due to concerns about costs and the disruption of public spaces during installation.

The city focused on transparency and inclusivity by involving community members in planning. Public forums, surveys, and interactive design sessions allowed residents to voice their opinions and contribute to the project's development. Sydney also highlighted the long-term economic and environmental benefits, such as reduced maintenance costs and improved public health, to justify the initial investment. This participatory approach helped build trust and support for the project.

Chapter 9: Future Trends and Innovations

As urban areas grapple with the challenges posed by climate change, developing and implementing innovative solutions such as cool pavements become increasingly important. Cool pavements, designed to mitigate the UHI effect, improve air quality, and enhance urban sustainability, are evolving rapidly thanks to advancements in technology and research. This chapter explores the future trends and innovations in cool pavements, examining emerging technologies, promising research directions, and the potential for global adoption.

We begin by delving into the latest materials and technologies being developed for cool pavements. These advancements aim to enhance cool pavements' performance, durability, and cost-effectiveness, making them more attractive and feasible for widespread use. We will explore cutting-edge innovations that promise to revolutionize how cities manage heat and water through their infrastructure.

Next, we discuss future research directions to drive the next generation of cool pavements. Identifying key areas for further investigation, this section highlights the importance of collaboration between academic institutions, industry, and government agencies. By focusing on interdisciplinary research and development, we can accelerate the adoption of cool pavements and maximize their benefits.

Finally, we examine the potential for global adoption of cool pavement solutions. Scaling up these innovations to a global level presents unique challenges and opportunities. Through global case studies and best practices, we will illustrate how different cities worldwide are successfully implementing cool pavements and overcoming barriers to adoption. By learning from these examples, other cities can develop strategies to integrate cool pavements into their urban planning and climate resilience efforts.

In conclusion, this chapter aims to provide a comprehensive overview of the future trends and innovations in cool pavements, offering insights into how these advancements can contribute to creating more sustainable and resilient urban environments. By embracing emerging technologies, fostering research and collaboration, and promoting global adoption, we can unlock the full potential of cool pavements to address the pressing challenges of urbanization and climate change.

Emerging Technologies in Cool Pavements

Emerging technologies in cool pavements are transforming how cities address urban heat and sustainability challenges, offering innovative materials and solutions that enhance performance and durability.

New Materials and Technologies

The development of new materials and technologies is driving significant advancements in the field of cool pavements, enabling cities to better manage urban heat and improve sustainability. These innovations focus on enhancing pavement materials' reflectivity, durability, and permeability, making them more effective at mitigating the UHI effect and improving stormwater management.

Reflective Coatings

Reflective coatings are one of the most widely adopted technologies in cool pavements. These coatings are designed to increase pavement surfaces' albedo, or reflectivity, thereby reducing the amount of heat absorbed. Traditional asphalt pavements can have an albedo as low as 0.05 to 0.10, absorbing 90-95% of incoming sunlight. Reflective coatings can increase albedo to 0.30 or higher, significantly reducing surface temperatures.

Recent advancements in reflective coatings include developing materials that maintain high reflectivity over time, even in harsh weather conditions. These coatings are formulated to resist fading, dirt accumulation, and wear, ensuring long-term effectiveness. Additionally, color-tunable reflective coatings allow for aesthetic customization without compromising reflectivity, making them suitable for various urban environments.

High-Albedo Aggregates

Another promising technology involves the use of high-albedo aggregates in the production of concrete and asphalt pavements. These aggregates, typically made from light-colored materials such as quartz or limestone, can be mixed into the pavement to increase its reflectivity.

Using recycled materials, such as crushed glass and ceramic tiles, as high-albedo aggregates is gaining traction. These recycled materials enhance reflectivity and contribute to sustainability by diverting waste from landfills. Research is ongoing to optimize the mix designs and ensure the structural integrity of pavements incorporating these innovative aggregates.

Permeable Pavements

Permeable pavements, including porous asphalt, pervious concrete, and permeable interlocking concrete pavers, allow water to infiltrate the surface, reducing surface runoff and promoting groundwater recharge. These pavements are particularly effective in managing stormwater and reducing flooding risks in urban areas.

Recent developments in permeable pavement technology focus on improving their load-bearing capacity and durability. For instance, incorporating geosynthetic materials and advanced bonding agents enhances the structural performance of permeable pavements, making them suitable for a wider range of applications, including high-traffic areas. New designs that combine permeability with high reflectivity are also being explored to provide dual benefits of cooling and stormwater management.

Phase Change Materials (PCMs)

Phase change materials (PCMs) are an emerging technology that can be integrated into cool pavements to enhance thermal performance. PCMs absorb and store heat during the day by changing from solid to liquid and release the stored heat at night by reverting to a solid state. This process helps to regulate pavement temperatures and reduce peak surface heat.

Research is focused on developing PCMs that are stable, non-toxic, and capable of withstanding the mechanical stresses of pavement applications. Encapsulating PCMs within durable shells and incorporating them into pavement materials such as concrete and asphalt are key areas of exploration. These innovations aim to maximize the thermal benefits of PCMs while ensuring the longevity and structural integrity of the pavements.

Photocatalytic Pavements

Photocatalytic pavements are another innovative technology that helps to cool urban areas and improve air quality. These pavements are treated with photocatalytic compounds, such as titanium dioxide, which catalyze the breakdown of pollutants like nitrogen oxides (NOx) and volatile organic compounds (VOCs) when exposed to sunlight.

Advances in photocatalytic technology include the development of more efficient and long-lasting compounds that can be integrated into various pavement types. Research is also being conducted on optimizing the surface texture of photocatalytic pavements to enhance their pollutant removal efficiency while maintaining durability and skid resistance.

Future Prospects and Innovations

The future of cool pavements is promising, with ongoing research and technological advancements paving the way for more effective and versatile solutions. As urban areas continue to grow and face the challenges of climate change, the need for innovative pavement technologies that can mitigate the UHI effect, manage stormwater, and enhance urban sustainability becomes increasingly critical. This section explores the future prospects and innovations likely to shape the next generation of cool pavements.

Integration with Smart Technologies

One of the most exciting prospects for cool pavements is their integration with smart technologies. Smart pavements equipped with sensors and Internet of Things (IoT) capabilities can provide real-time data on temperature, moisture levels, and structural integrity. This data can be used to monitor the performance of cool pavements and optimize their maintenance and management.

Future developments may include self-monitoring pavements that can detect and repair minor cracks autonomously, extending their lifespan and reducing maintenance costs. Additionally, smart pavements can be integrated with urban cooling systems, such as water misting or shading devices, to enhance their effectiveness in reducing urban temperatures.

Advanced Materials

Developing advanced materials is another area with significant potential for innovation in cool pavements. Researchers are exploring using nanotechnology and advanced composites to create materials with superior reflectivity, durability, and environmental benefits.

Nano-engineered coatings that offer enhanced UV resistance and self-cleaning properties are being developed to maintain high reflectivity over time. These coatings can reduce maintenance requirements and extend the functional life of cool pavements. Additionally, incorporating recycled and bio-based materials in cool pavements can improve their sustainability profile and reduce environmental impact.

Multifunctional Pavements

The future of cool pavements also lies in creating multifunctional surfaces that offer multiple environmental and social benefits. These pavements can combine cooling properties with other functions, such as energy generation, pollutant removal, and aesthetic enhancements.

Solar pavements that generate electricity from sunlight while keeping urban areas cool are an emerging technology with great potential. These pavements can be integrated with energy storage systems to provide renewable energy for street lighting and other urban infrastructure. Similarly, pavements incorporating photocatalytic materials to break down pollutants can improve air quality and reduce surface temperatures.

Modular and Flexible Designs

Modular and flexible pavement designs are gaining attention as a way to facilitate the installation and maintenance of cool pavements. These designs allow for easier replacing damaged sections and adaptability to different urban environments.

Interlocking pavement systems that can be easily assembled and disassembled offer the flexibility needed for rapid deployment and maintenance. These systems can be customized to fit various urban settings, from pedestrian walkways to busy streets, enhancing their versatility and practicality.

Enhanced Public Awareness and Adoption

As the benefits of cool pavements become more widely recognized, efforts to increase public awareness and adoption are expected to grow. Education campaigns and community engagement initiatives can help build support for cool pavement projects and encourage their inclusion in urban planning and development.

Future prospects include the development of interactive platforms and tools that educate the public about the benefits of cool pavements and how they contribute to urban sustainability. Virtual reality (VR) and augmented reality (AR) applications can provide immersive experiences that showcase cool pavements' impact in real-world scenarios, helping gain public support and drive adoption.

Future Research Directions

Future research directions in cool pavements focus on exploring new materials, improving existing technologies, and understanding the long-term impacts of these innovations on urban environments.

Areas for Further Research

As the field of cool pavements continues to evolve, several key areas require further research to enhance their effectiveness, durability, and overall impact on urban sustainability. By addressing these research gaps, we can develop more advanced cool pavement solutions that better meet the needs of modern cities facing the challenges of climate change and urbanization.

Long-Term Performance and Durability

One critical area for further research is cool pavements' long-term performance and durability. While many cool pavement technologies show promising results in the short term, their effectiveness over extended periods remains a concern. Factors such as wear and tear from traffic, weathering, and the accumulation of dirt and debris can impact the performance of cool pavements.

Studies should focus on understanding how different cool pavement materials and coatings perform under various environmental conditions over time. This includes investigating the effects of traffic load, climate, and maintenance practices on the longevity and reflectivity of cool pavements. Developing more durable materials and maintenance protocols can help ensure that cool pavements remain effective in the long term.

Cost-Benefit Analysis

Conducting comprehensive cost-benefit analyses of cool pavement technologies is essential to determine their economic feasibility and justify their adoption on a larger scale. While cool pavements' environmental and social benefits are well-documented, a detailed understanding of the financial implications is needed.

Future research should include detailed economic assessments considering the initial costs, maintenance expenses, energy savings, and health benefits associated with cool pavements. Comparing these costs and benefits with those of traditional pavements can provide valuable insights for policymakers and urban planners, facilitating informed decision-making.

Integration with Other Sustainable Technologies

Another important area for research is the integration of cool pavements with other sustainable technologies and infrastructure. Combining cool pavements with green roofs, urban forestry, and renewable energy systems can create synergistic effects that enhance overall urban sustainability.

Investigating how cool pavements can be effectively integrated with other green infrastructure elements can help maximize their benefits. Research should explore the potential for creating multifunctional surfaces that simultaneously provide cooling, stormwater management, energy generation, and pollutant reduction. Additionally, studying the interactions between cool pavements and urban ecosystems can provide insights into optimizing their design and implementation.

Impact on Urban Microclimates

Understanding cool pavements' impact on urban microclimates is crucial for assessing their effectiveness in mitigating the UHI effect. While cool pavements can reduce surface temperatures, their influence on ambient air temperatures and overall urban heat dynamics needs further exploration.

Advanced modeling and field studies should be conducted to assess cool pavements' impact on local microclimates. This includes examining how cool pavements affect air temperature, humidity, wind patterns, and thermal comfort at different scales, from individual streets to entire neighborhoods. Such studies can help optimize the placement and design of cool pavements to maximize their cooling benefits.

Collaboration Opportunities

Collaboration is key to advancing the development and implementation of cool pavements. By bringing together diverse expertise and resources, stakeholders can drive innovation, overcome challenges, and maximize the benefits of cool pavement technologies. Various collaboration opportunities exist across academia, industry, government, and community organizations.

Academic and Research Institutions

Academic and research institutions are crucial in advancing the science and technology behind cool pavements. Collaborations between universities, research centers, and government agencies can lead to significant material science, engineering, and environmental impact assessment breakthroughs.

Joint research projects and grants can be established to study cool pavements' long-term performance, durability, and environmental benefits. Interdisciplinary research involving urban planning, civil engineering, environmental science, and economics can provide a holistic understanding of cool pavements and inform policy decisions. Additionally, academic partnerships can facilitate the development of advanced modeling tools to predict the impacts of cool pavements on urban microclimates and energy consumption.

Industry Partnerships

Industry partnerships are vital for translating research findings into practical applications and scaling up cool pavement technologies. Companies specializing in construction materials, pavement technologies, and urban infrastructure can collaborate with researchers and municipalities to develop, test, and deploy innovative cool pavement solutions.

Collaborative efforts can focus on developing new materials and coatings that enhance the reflectivity and durability of cool pavements. Industry partners can also contribute to pilot projects demonstrating cool pavements' effectiveness in real-world settings. These projects can provide valuable data and insights, helping to refine technologies and improve implementation strategies. Additionally, industry collaborations can support the development of standards and guidelines for widely adopting cool pavements.

Government and Policy Makers

Government agencies and policymakers play a crucial role in creating the regulatory and funding environment needed to successfully implement cool pavements. Collaboration between different levels of government and other stakeholders can facilitate the integration of cool pavements into urban planning and sustainability initiatives.

Governments can work with academic and industry partners to develop policies, incentives, and funding programs that support cool pavement projects. PPPs can leverage resources from both sectors to finance and implement large-scale projects. Collaborative efforts can also focus on developing training and capacity-building programs for municipal staff and contractors, ensuring they have the knowledge and skills needed to implement cool pavements effectively.

Community Organizations and Public Engagement

Community organizations and public engagement are essential for building support and ensuring the successful adoption of cool pavements. Engaging residents and local groups can help address concerns, gather valuable input, and foster a sense of ownership and commitment to the project.

Collaboration with community organizations can involve educational campaigns, public meetings, and participatory planning processes that involve residents in decision-making. These efforts can help raise awareness about the benefits of cool pavements and encourage community support. Additionally, partnerships with schools and local businesses can create opportunities for hands-on learning and demonstration projects that showcase the positive impacts of cool pavements on urban environments.

Potential for Global Adoption

The potential for global adoption of cool pavements is immense. They offer a sustainable solution to urban heat and environmental challenges faced by cities worldwide.

Scaling Up Cool Pavement Solutions

Scaling up cool pavement solutions from pilot projects to widespread implementation requires strategic planning, substantial investment, and strong collaboration among various stakeholders. The transition from small-scale demonstrations to large-scale adoption involves addressing several key factors to ensure that cool pavements become an integral part of urban infrastructure globally.

Policy and Regulatory Support

One critical step in scaling up cool pavement solutions is securing robust policy and regulatory support. Local, regional, and national governments need to create favorable policies that mandate or incentivize the use of cool pavements in urban development projects. This includes updating building codes, implementing zoning regulations requiring reflective or permeable pavements, and providing tax credits or grants for projects incorporating cool pavement technologies.

Municipalities can develop comprehensive urban heat management plans that prioritize cool pavements as a key strategy. National governments can support these efforts by funding research and development, pilot projects, and large-scale implementations.

Funding and Financial Incentives

Securing adequate funding and developing financial incentives are crucial for adopting cool pavements. Initial costs can be a barrier, especially for cities with limited budgets. However, long-term savings in energy costs, reduced maintenance expenses, and public health benefits can justify the investment.

Governments and financial institutions can offer green bonds, low-interest loans, and PPP models to fund cool pavement projects. Establishing dedicated funds or grants for urban sustainability projects can encourage municipalities to adopt cool pavements.

Public Awareness and Education

Educating the public and raising awareness about the benefits of cool pavements is essential for gaining widespread support. Community engagement initiatives can help residents understand how cool pavements reduce urban heat, improve air quality, and enhance the overall livability of their neighborhoods.

Cities can launch public awareness campaigns, conduct workshops, and provide informational materials to educate residents and businesses about the advantages of cool pavements. Engaging local media and using social media platforms can further amplify these efforts.

Technical Training and Capacity Building

Ensuring sufficient technical expertise to implement and maintain cool pavements is vital for their successful adoption. Training programs for urban planners, engineers, and construction workers can help build the necessary skills and knowledge.

Governments and educational institutions can develop certification programs and offer workshops focused on cool pavement technologies. Collaboration with industry experts and research institutions can provide ongoing professional development opportunities.

Chapter 10: Conclusion

The conclusion of this guide on cool pavements underscores the critical importance of innovative urban infrastructure solutions in addressing the challenges of climate change and urbanization.

Summary of Key Points

Recap of Major Findings and Insights

Throughout this guide, we have explored the significant potential of cool pavements in mitigating the UHI effect, improving urban sustainability, and enhancing the quality of life in cities. We began by understanding the UHI phenomenon and its adverse impacts on urban environments, such as increased temperatures, higher energy consumption, and deteriorating air quality. Cool pavements emerged as an effective solution to these challenges, with their ability to reflect more sunlight, absorb less heat, and facilitate better stormwater management.

We examined various cool pavements, including reflective, permeable, and vegetated options, each offering unique benefits and applications. Reflective pavements, with their high albedo materials, significantly reduce surface temperatures, while permeable pavements improve water infiltration and reduce flooding risks. Vegetated pavements, integrating plant life into the urban fabric, provide additional cooling and aesthetic benefits.

Key case studies demonstrated the successful implementation of cool pavements in cities like Los Angeles, Chicago, Phoenix, and Sydney. These examples highlighted the importance of pilot projects, community engagement, and strategic partnerships in overcoming technical, financial, and institutional barriers. Moreover, we discussed emerging technologies and future innovations that promise to enhance the performance and applicability of cool pavements, such as phase change materials, photocatalytic compounds, and smart pavements with IoT capabilities.

Importance of UHI Mitigation

Mitigating the Urban Heat Island effect is crucial for creating sustainable, livable cities in the face of climate change. UHI exacerbates the impacts of heatwaves, increases energy demand, and contributes to poor air quality, all of which severely affect public health and urban ecosystems. By reducing urban temperatures, cool pavements play a vital role in addressing these issues and promoting environmental justice.

Cool pavements contribute to significant energy savings by lowering the need for air conditioning, especially during peak summer months. This reduction in energy consumption decreases greenhouse gas emissions and alleviates the strain on power grids, enhancing the resilience of urban energy systems. Furthermore, by improving air quality through reduced temperatures and pollutant removal technologies, cool pavements help combat respiratory and cardiovascular diseases, particularly in vulnerable populations.

Another critical benefit of cool pavements is effective stormwater management. Permeable cool pavements reduce surface runoff, prevent flooding, and recharge groundwater supplies, essential for maintaining urban water systems' health. These benefits are particularly important as climate change increases the frequency and intensity of extreme weather events.

In summary, mitigating the UHI effect through cool pavements is essential for building resilient, sustainable cities. By addressing the multifaceted challenges urban heat poses, cool pavements enhance public health, environmental quality, and overall urban livability. As we look to the future, the continued development and implementation of cool pavements will be a cornerstone of sustainable urban planning and climate adaptation strategies.

Call to Action for Urban Planners and Policymakers

Steps to Promote Cool Pavements

Urban planners and policymakers play a crucial role in promoting the adoption and implementation of cool pavements in cities. To effectively integrate cool pavements into urban infrastructure, several key steps should be taken:

1. Develop Comprehensive Plans: Create urban heat management plans that prioritize using cool pavements. These plans should outline specific goals, timelines, and strategies for reducing urban heat and improving sustainability through cool pavement projects.
2. 2. Conduct Pilot Projects: Implement pilot projects to demonstrate the effectiveness of cool pavements in various urban settings. These projects can serve as proof-of-concept and provide valuable data on performance, durability, and public acceptance. Use the findings from pilot projects to refine designs and implementation strategies.
3. Secure Funding and Resources: Identify and secure funding sources to support cool pavement initiatives. This can include applying for grants, issuing green bonds, or developing PPPs. Allocate budget resources specifically for maintaining and monitoring cool pavements to ensure their long-term effectiveness.
4. Engage Stakeholders: Actively involve community members, businesses, and other stakeholders in the planning and implementation. Conduct public meetings, workshops, and surveys to gather input and build support. Transparent communication and inclusive decision-making can help address concerns and foster a sense of ownership among stakeholders.
5. Provide Education and Training: Offer training programs for urban planners, engineers, and construction professionals to build the technical expertise needed for cool pavement projects. Educate the public about the benefits of cool

pavements through informational campaigns, school programs, and community events.

Recommendations for Policy and Practice

To ensure the successful adoption of cool pavements, urban planners and policymakers should implement the following recommendations for policy and practice:

- Update Building Codes and Zoning Regulations: Incorporate requirements for cool pavements into building codes and zoning regulations. This can include mandates for reflective or permeable pavements in new developments and incentives for retrofitting existing infrastructure.
- Offer Financial Incentives: Develop financial incentives to encourage using cool pavements. This can include tax credits, rebates, or grants for property owners and developers who incorporate cool pavements into their projects. Financial incentives can help offset the initial costs and promote widespread adoption.
- Establish Standards and Guidelines: Create clear standards and guidelines for designing, installing, and maintaining cool pavements. These guidelines should specify material specifications, performance criteria, and best practices to ensure their effectiveness and durability.
- Monitor and Evaluate: Implement monitoring and evaluation frameworks to track the performance of cool pavement projects. Collect data on surface temperatures, energy savings, stormwater management, and public health impacts. Use this data to continuously improve designs and practices.
- Foster Collaboration: Encourage collaboration between government agencies, academic institutions, industry partners, and community organizations. Collaborative efforts can drive innovation, share resources, and ensure that cool pavement projects are well-coordinated and aligned with broader sustainability goals.
- Advocate for Federal and State Support: Advocate for federal and state policies that support urban heat mitigation and cool

pavement initiatives. This can include lobbying for funding, research support, and regulatory frameworks that promote the use of cool pavements.

Vision for the Future of Cool Pavements in Sustainable Cities

Future Outlook and Aspirations

The future outlook for cool pavements is both promising and transformative. As cities grapple with the effects of climate change and rapid urbanization, cool pavements are poised to become a cornerstone of sustainable urban infrastructure. Looking ahead, we can expect significant advancements in cool pavement technologies, driven by ongoing research and innovation. These advancements will enhance cool pavements' performance, durability, and cost-effectiveness, making them an increasingly attractive option for cities worldwide.

Our aspirations for the future of cool pavements include widespread adoption and integration into urban planning and development practices. We can ensure that new developments and major renovations contribute to urban cooling and sustainability goals by embedding cool pavements into city design guidelines and building codes. Additionally, implementing large-scale cool pavement projects across diverse urban environments will demonstrate their versatility and effectiveness in different climatic conditions.

In the future, we envision cool pavements mitigating the Urban Heat Island (UHI) effect and contributing to broader urban resilience strategies. As part of a holistic approach to climate adaptation, cool pavements will work in synergy with other green infrastructure elements, such as green roofs, urban forests, and renewable energy systems, to create more resilient and sustainable cities. Furthermore, integrating smart technologies will enable real-time monitoring and maintenance, optimizing the performance and longevity of cool pavements.

Role of Cool Pavements in Urban Sustainability

Cool pavements play a crucial role in advancing urban sustainability by addressing several key challenges modern cities face. One of the most significant contributions of cool pavements is their ability to reduce urban temperatures, thereby mitigating the UHI effect. Lower surface and ambient temperatures decrease energy demand for cooling buildings, leading to substantial energy savings and reduced greenhouse gas emissions. This reduction in energy consumption supports climate mitigation efforts and enhances urban energy systems' resilience.

In addition to cooling benefits, cool pavements contribute to improved air quality by reducing the formation of ground-level ozone and other pollutants. This has significant public health implications, particularly for vulnerable populations such as children, the elderly, and individuals with respiratory conditions. By promoting cleaner air, cool pavements help create healthier urban environments.

Cool pavements also enhance urban sustainability through effective stormwater management. Permeable cool pavements facilitate water infiltration, reducing surface runoff and mitigating the risk of flooding. This natural infiltration process helps recharge groundwater supplies and maintain the health of urban water systems. By managing stormwater more sustainably, cool pavements reduce the burden on municipal drainage systems and prevent water pollution.

Furthermore, cool pavements contribute to social equity by improving the livability of urban areas. Cooler, more comfortable public spaces encourage outdoor activities and social interactions, fostering a sense of community and well-being. By prioritizing the implementation of cool pavements in underserved neighborhoods, cities can address environmental justice issues and ensure that all residents benefit from these innovative solutions.

In conclusion, cool pavements are a vital component of sustainable urban development. Their ability to mitigate urban heat, improve air quality, manage stormwater, and enhance social equity underscores their importance in creating resilient, livable cities. As we look to the future, the continued advancement and widespread adoption of cool pavements will play a pivotal role in shaping sustainable urban environments that are prepared to meet the challenges of climate change and urbanization.

www.ingramcontent.com/pod-product-compliance
Lightning Source LLC
Chambersburg PA
CBHW050104230526
45470CB00004B/1669